iPad 7th Generation

The Beginners, Dummies and Seniors Guide to

Maximizing Your iPad

(The User Manual like No Other)

4th Edition

Phila Perry

978-1-63750-023-1

Published By:

The User Manual Like No Other

Website: www.engolee.com/bookdeals

Email: info@engolee.com

Table of Contents

FREE BONUS

Are you passionate about reading and adding new value to yourself? Would you love to get various **FREE & DISCOUNTED** books delivered to your inbox

GET ACCESS NOW!

or

Visit http://www.engolee.com/bookdeals

Introduction

This book is a guide for all of Apples iPad models such as iPad 2,iPad (3rd generation), iPad (4th generation), iPad Air, iPad Air 2, iPad Pro (12.9-inch), iPad Pro (9.7-inch), iPad (7th generation), iPad Pro (12.9-inch) (2nd generation), iPad Pro (10.5-inch), iPad (6th generation), iPad Pro (11-inch), iPad Pro (12.9-inch) (3rd generation), iPad Air (3rd generation) features exclusively. It includes everything from necessary setup information to finding and installing new apps to using the iPad Pro for communication, entertainment, and productivity.

The information presented in this book is targeted at kids, teens, adolescents, and adults who are probably a beginner or dummies, seniors, or experts with the use of iPad tablets in a simple to understand and follow steps.

In this book, you will find Step-by-step instructions including how to fix some technical iPad Pro problems in simple terms.

The book is easy, clear, readable, and focuses on what you want to do with your iPad tablet such as:

- Secret features unravelled in the original manual
- Personalizing the feel and look of your iPad
- fixing slow iPad problems yourself
- Screen splitting of your iPad device
- How to activate used iPad
- Use Siri's voice commands to control your iPad and for other exclusive things you never would have imagined.
- Steps for fixing iPad that won't charge or power ON
- How to extends iPad's battery strength
- Use iCloud to store and share your photos and other essential data online

• Troubleshoot common iPad problems

...and many more.

This is the book to learn how to get the most out of your iPad Pro Now!

CHAPTER 1

iPad First Time Set-Up

There is no need for connecting your brand-new iPad to your personal computer, as long as there is a mobile data connection designed for activation. As you end the set-up wizard, you may navigate back by tapping the back arrow at the top left-hand side of the screen and scroll further to another display by tapping another button at the top right-hand corner.

You can commence by pressing down the power button at the top edge of your brand-new iPad. You may want to keep it pressed down for about two seconds until you notice a vibration, meaning the iPad is booting up.

Once it boots up finally, you can start initial set up by

following the processes below;

- Swipe your finger over the display screen to start the set-up wizard.
- Choose the language of preference - English is usually at the top of the list, so there is no problem finding it. However, if you would like to apply a different language, scroll down to look for your desired language, and tap to select the preferred language.
- Choose your country - United States may be near the top of the list. If otherwise, scroll down the list and select the United States or any of your choice.
- You need to connect your iPad to the internet to start its activation. You can test this via a link with a Wi-Fi network. Locate the name of your available network in the list shown, and then tap

on it to select it.

- Enter the Wi-Fi security password (you will generally find this written on your router, which is probably known as the WPA Key, WEP Key, or Password) and select Sign up. A tick indication shows you are connected, and a radio image appears near the top of the screen. The iPad will now start activation with Apple automatically. This may take some time!
- In case your iPad is a 4G version, you will be requested to check for updated internet configurations after inserting a new Sim card. You can test this anytime, so, for the present time, tap **Continue**.
- Location services will help you with mapping, weather applications, and more, giving you specific information centered wholly on what your

location is. Select whether to use location service by tapping allow location services.

- You would now be requested to create **Touch ID,** which is Apple's fingerprint identification. **Touch ID** allows you to unlock your iPad with your fingerprint instead of your passcode or security password. To set up Tap Identification, put a finger or your thumb on the home button (but do not press it down!). To by-pass this for the moment, tap ***setup Tap Identification later***.
- If you are establishing Touch ID, the tutorial instruction on the screen will walk you through the set-up process. Put your finger on the home button, then remove it till the iPad has properly scanned your fingerprint. Whenever your print is wholly scanned, you will notice a screen letting you know that tap recognition is successful. Tap **Continue**.

- You will be requested to enter a passcode to secure your iPad. If you create **Touch ID**, you must use a passcode if in any case your fingerprint isn't acknowledged. Securing your computer data is an excellent idea, and the iPad provides you with several options. Tap password option to choose your lock method.
- You can arrange a Custom Alphanumeric Code (that is a security password that uses characters and figures), a Custom Numeric Code (digit mainly useful, however, you can add as many numbers as you want!) or a 4-Digit Numeric Code (a high old college pin!). In case you didn't install or setup **Touch ID** you may even have an option not to add Security password. Tap on your selected Security option.
- I would recommend establishing a 4-digit numeric

code, or Touch ID for security reasons but all optional setup is done likewise. Input your selected Security password using the keyboard.

- Verify your Security password by inputting it again. If the Password does not match, you'll be requested to repeat! If indeed they do match, you'll continue to another display automatically.

At this time of the set-up process, you'll be asked whether you have used an iPad before and probably upgrading it, you can restore all of your applications and information from an iCloud or iTunes backup by deciding on the best option. If this is your first iPad, you will have to get it started as new, yet, in case you are moving from Android to an iPad, you can transfer all your data by deciding and choosing the choice you want.

How to Restore iPad Back-up from iCloud or iTunes

If you want to restore your iPad from an iTunes back-up, you may want to connect to iCloud and have the latest version of iTunes installed on it. If you are ready to begin this process, tap **restore** from iTunes back-up on your iPad and connect it to your personal computer. Instructions about how to bring back your data can be followed on the laptop screen.

In case your old iPad was supported on iCloud, then follow the instructions below to restore your applications & data to your brand-new device:

- Tap ***Restore*** from iCloud back-up.
- Register with the **Apple ID** and Password that you applied to your old iPad. If you fail to recollect the security password, there's a link that may help you

reset it.

- The Terms & Conditions screen will show. Tap the links to learn about specific areas in detail. When you are ready to proceed, select **Agree**.
- Your **iPad** will need some moments to create your **Apple ID** and hook up with the **iCloud server**.
- You will notice a summary of available backups to download. The most up-to-date backup will be observed at the very top, with almost every other reserve below it. If you want to restore from a desirable backup, tap the screen for ***all backups*** to see the available choices.
- Tap on the back-up you want to restore to start installing.
- A progress bar will be shown, providing you with a demo of the advancement of the download.

When the restore is completed, the device will restart.

- You would see a notification telling you that your iPad is updated effectively. Tap ***Continue***.
- To complete the iCloud set up on your recently restored iPad, you should re-enter your iCloud (**Apple ID**) password. Enter/review it and then tap ***Next***.
- You'll be prompted to upgrade the security information related to your ***Apple ID***. Tap on any stage to replace your computer data, or even to bypass this option. If you aren't ready to do this, then tap the ***Next*** button.
- **Apple pay** is Apple's secure payment system that stores encrypted credit or debit card data on your device and making use of your iPad also with your fingerprint to make safe transactions online and

with other apps. Select *Next* to continue.

- To ***feature/add a card***, place it on a set surface and place the iPad over it, so the card is put in the camera framework. The credit card info will be scanned automatically, and you will be requested to verify that the details on display correspond with your card. You'll also be asked to enter the ***CVV*** (safety code) from the personal strip behind the card. If you choose (or the camera cannot recognize your cards), you can enter credit card information by hand by tapping the hyperlink. You could bypass establishing **Apple Pay** by tapping ***create later***.
- Another screen discusses the ***iCloud keychain***, which is Apple's secure approach to sharing your preserved security password and payment information throughout all your Apple devices.

You might use *iCloud security code* to validate your brand-new device and import present data, or you might be asked to continue registering your keychain if it's your first Apple device. In case you don't want to share vital data with other devices, you should go to *avoid iCloud keychain* or *don't restore passwords*.

- If you selected to set up your **Apple keychain**, you'll be notified to either use a Security password (the same one you'd set up on your iPad) or produce a different code. If you're making use of your iCloud security code, you should put it on your iPad when prompted.
- This will confirm your ID when signing on to an iCloud safety code; a confirmation code will be delivered via SMS. You may want to hyperlink your smartphone text code (if you have never

distributed one with Apple already) so that the code may be provided as a text. Then enter this code to your iPad if requested, then select ***Next.***

- You'll then be asked to create **Siri**. ***Siri*** is your own digital personal associate, which might search the internet, send communications, and check out data in your device and a lot more, all without having to flick via specific apps. Choose to create Siri by tapping the choice or start Siri later to skip this task for now.
- To set up and create **SIRI**, you will need to speak several phrases to the iPad to review your conversation patterns and identify your voice.
- Once you say every term, a tick will be observed, showing that it's been known and comprehended. Another phrase may indicate that you should read aloud.

- Once you've completed the five phrases, you will notice a display notifying that Siri has been set up correctly. Tap ***Continue***.
- The iPad display alters the color balance to help make the screen show up naturally under distinctive light conditions. You can switch this off in the screen settings after the iPad has completed configuring it. Tap ***continue*** to continue with the setup.
- Has your iPad been restored? Tap begin to transfer your computer data to your brand-new iPad.
- You'll be prompted to ensure your brand-new iPad has enough power to avoid the device turning off in the process of downloading applications and information. Tap ***OK*** to verify this recommendation.
- You will notice a notification show up on your

apps, to download in the background.

How to Move Data From Android

Apple has made it quite easy to move your data from a Google Android device to your new iPad. Proceed to the iOS app. I'll direct you about how to use the application to move your data!

- Using the iPad, if you are on the applications & data screen of the set-up wizard, tap ***move data from Google android***.
- Go to the Play Store on your Google android device and download the app recommended by the set-up wizard. When it is installed, open up the app, select **Continue** and you'll be shown the ***Terms & Conditions*** to continue.
- On your Android device, tap ***Next*** to start linking

your Devices. On your own iPad, select ***Continue***.

- Your iPad will show a 6-digit code which has to be received into the Google android device to set the two phone up.
- Your Google android device will screen all the data that'll be moved. By default, all options are ticked - so if there could be something you don't want to move, tap the related collection to deselect it. If you are prepared to continue, tap ***Next*** on your Google android device.
- As the change progresses, you will notice the iPad display screen changes, showing you the position of the info transfer and progress report.
- When the transfer is completed, you will notice a confirmation screen on each device. On your Android Device, select ***Done*** to shut the app. On your iPad, tap ***Continue Installing iPad***.

- An **Apple ID** allows you to download apps, supported by your iPad and synchronize data through multiple devices, which makes it an essential account you should have on your iPad! If you have been using an iPad previously, or use iTunes to download music to your laptop, then you should have already become an **Apple ID user**. Register with your username and passwords (when you have lost or forgotten your **Apple ID** or password you will see a link that may help you reset it). If you're not used to iPad, select doesn't have an Apple ID to create one for free.
- The Terms & Conditions for your iPad can be seen. Please go through them (tapping on more to study additional info), so when you are done, tap ***Agree***.
- You'll be asked about synchronizing your data with iCloud. That's to ensure bookmarks,

connections and other items of data are supported securely with your other iPad's data. Tap ***merge*** to permit this or ***don't merge*** if you'll have a choice to keep your details elsewhere asides iCloud.

- **Apple pay** is Apple's secure payment system that stores encrypted credit or debit card data on your device and making use of your iPad also with your fingerprint to make safe transactions online and with other apps. Select ***Next*** to continue.
- To ***feature/add a card***, place it on a set surface and place the iPad over it, so the card is put in the camera framework. The credit card info will be scanned automatically, and you will be requested to verify that the details on display correspond with your card. You'll also be asked to enter the ***CVV*** (safety code) from the personal strip behind the card. If you choose (or the camera cannot

recognize your cards), you can enter credit card information by hand by tapping the hyperlink. You could bypass establishing **Apple Pay** by tapping ***create later***.

- Another screen discusses the ***iCloud keychain***, which is Apple's secure approach to sharing your preserved security password and payment information throughout all your Apple devices. You might use *iCloud security code* to validate your brand-new device and import present data, or you might be asked to continue registering your keychain if it's your first Apple device. In case you don't want to share vital data with other devices, you should go to *avoid iCloud keychain* or *don't restore passwords*.
- If you selected to set up your Apple keychain, you'll be notified to either use Security password

(the same one you'd set up on your iPad) or produce a different code. If you're making use of your iCloud security code, you should put it on your iPad when prompted.

- This will confirm your ID when signing on to an iCloud safety code; a confirmation code will be delivered via SMS. You may want to hyperlink your smartphone text code (if you have never distributed one with Apple already) so that the code may be provided as a text. Then enter this code to your iPad if requested, then select ***Next.***
- You'll then be asked to create **Siri**. ***Siri*** is your own digital personal associate, which might search the internet, send communications, and check out data in your device and a lot more, all without having to flick via specific apps. Choose to create Siri by tapping the choice or start Siri later to skip

this task for now.

- To set up and create SIRI, you will need to speak several phrases to the iPad to review your conversation patterns and identify your voice.
- Once you say every term, a tick will be observed, showing that it's been known and comprehended. Another phrase may indicate that you should read aloud.
- Once you've completed the five phrases, you will notice a display notifying that Siri has been set up correctly. Tap ***Continue***.
- The iPad display alters the color balance to help make the screen show up naturally under distinctive light conditions. You can switch this off in the screen settings after the iPad has completed configuring it. Tap ***continue*** to continue with the

setup.

- Has your iPad been restored? Tap begin to transfer your computer data to your brand-new iPad.
- You'll be prompted to ensure your brand-new iPad has enough power to avoid the device turning off in the process of downloading applications and information. Tap ***OK*** to verify this recommendation.
- You will notice a notification show up on your apps, to download in the background.

NB: *Setting up as a new iPad: Similar method, as described above, applies.*

CHAPTER 2

How to Add Email Account(s)

The iPad allows multiple POP3, IMAP, and other accounts. If you have one email, take into account one for work and another one for private/home, getting all of your emails is as simple as adding every of the account on your iPad.

- To commence, select the *Configurations icon*.
- In the configurations, find and tap on Email, on another page, tap on Account to add.
- At the right-hand side, tap ***ADD ACCOUNT***.
- Choose your E-mail providers (e.g. Google, Yahoo, Live, etc…) from the list shown. Mail Accounts is often utilized by companies and network service providers, live.com is the new name for a Hotmail

or windows live accounts, and YAHOO!, GOOGLE and AOL are self-explanatory. For another e-mail company, select ***OTHER*** followed by tapping ***ADD Email Account***.

- Review your **EMAIL** and Security password (aside from some other data required, alongside your name) as asked, and Tap **Next** or **Register** on the screen to continue.
- If the account(s) is established, you may review your details was successfully founded, or a data screen about how exactly your e-mail accounts can be utilized. You might allow this or tap **Next** to continue.
- This may then automatically spread to show notifications to be able to pick and choose which data the accounts will use on your new iPad - email, connections, calendars, and probably

different items too, with regards to the accounts. ***Turn OFF*** notifications for just about any items you do not need to be on your iPad, then select **Save**.

- Once completed, you should see your accounts in the list at the right-hand part of the screen.

In case your e-mail account does not install or get created, you'll be requested for additional information regarding the account, which includes incoming email server settings. You might contact them simultaneously for help. Once you've introduced your accounts, you could gain access to your email messages by tapping the email icon on your home screen. When you have installed several electronic mail accounts, you might view all your emails in a single inbox by selecting the ***ALL INBOXES*** option from the mailbox section in the e-mail app.

Adding & Importing Contact to Your iPad

When you have contacts on your old smartphone or device that you'll like to import to your brand-new iPad, please don't be worried I'll guide you!

Using Apple's iCloud service, you'll be able to import storage space documents and synchronize contacts simultaneously to your iPad.

If you have formerly been using an iPad, then transferring your contacts would be more comfortable by using ***Apple's iCloud Online Sync Service.***

In case your old mobile device has been installed to use iCloud, then your use of the same Apple identity (ID) on your iPad will deliver your contacts, calendar, and other information right to your brand-new device with no need for further action.

If you don't have iCloud installed on your old iPad, then

it's the very first thing you will need to configure. Your old iPad should be associated with a Wi-Fi network so that people can reproduce or duplicate the info from your mobile phone to iCloud.

Do the following on your old iPad:

- Locate and tap the Configurations icon.
- Scroll down and select ***iCloud***.
- If you see your address at the very top line, this means you're authorized directly into iCloud on your old iPad. If not, subscribe with the same Apple ID that you've used for your brand-new iPad.
- Turn internet network **ON.**
- Select the choice to Merge your computer data with **iCloud**. This will add all of your contacts to iCloud.

When you have already passed the initial set up the stage

for your iPad, you might activate the iCloud services following precise instructions as stated above. When you're connected directly to iCloud, your contacts will start to download to your brand-new iPad immediately. If you have configured your old iPad to the iCloud service, all you have to do is select the option to use iCloud through the initial set up of your **iPad**, as well as your contacts that will automatically show up on these devices.

How to Import Contacts From A Google Android To iPad

Your Android device can export its contacts into a storage space file, the precise form of a written report which iCloud has is with the capacity of managing and absorbing. Once your links are in iCloud, it is only a matter of time expecting the info to complete synchronizing on your iPad. However, if you are uncertain how to actualize this stage, then examine the steps below to discover more!

If the contacts aren't on the Google account on your old Android device, we'll need to get them there so that you can transfer these to your iPad, so that you will need to focus on step one as described below. If the contacts are already in your Google accounts, you may ignore this step.

To migrate your contacts out of your old Android device to your Google accounts:

- On your Android phone, tap the Contacts icon on your home display, or within the programs list.
- Tap the menu key, both as a button below the screen with three lines or the display screen button at the top-right corner, with three dots icon.
- Tap **Import/Export**. Several Android phones need you to press ***More*** before you start to see the *import/export settings.*
- Tap ***Export to SDCARD***, or ***Export to Storage Space*** depending on your mobile phone.
- When exported and you're back viewing the contacts list, select the ***menu key*** again.
- Tap ***Import/export,*** as done in the third step.

- Tap import from **SDCARD** or ***import from Storage space*** depending on your mobile phone.
- If you are asked where to import the contacts to, tap Google or the Google E-mail address.
- Based on your specific phone, you'll be requested to choose which contacts to import. If so, pick all links. Your links will now be on your Google accounts!

  Given that your contacts are on your Google accounts, you will extract these details and stick it onto the iCloud accounts such that it synchronizes to your iPad.
- On your laptop, head to Google's contacts website and subscribe with your Google email and password.
- From your Google contact, near the top of your

contacts, press ***More***, and consequently ***Export***.

- Ensure that the all contacts radio field is ticked, in addition to memory cards format- Press ***Export*** to download your contacts on your computer.
- From your laptop's web browser, go to the iCloud website and register to make use of your Apple ID and Password.
- Go through the Contacts, a summary of all your contacts presently residing on your **iCloud account**.
- Press the configurations icon in the bottom-left part of the contacts page. This appears as though it's a cog or tool.
- From your menu which shows up, go through the import button, and navigate to your download folder. After picking your cards to import and Press Okay, your contacts will begin to show in

iCloud! Within a few minutes, your iPad will start to show the same contacts too.

How to Import Contacts From a Blackberry Phone

To control your Blackberry contacts, we first need to transfer these to your computer. To do this, you must first download and set up the blackberry laptop software. As the software is installed, adhere to the instructions below!

- Connect your Blackberry to your laptop by using a micro USB cable for Blackberry processing device computer software to comprehend and identify the smartphone.
- Select Organizer at the left-hand panel of the program and tick the contacts field.
- While requested to choose the Sync Path, choose

your laptop/computer only.

- Below contacts account, be sure home windows contacts is chosen. Press **OK** to continue.
- Press sync organizer at the right-hand part of the program to move your Blackberry contacts onto your home windows address book.
- Given that your contacts are saved in windows contacts, you can synchronize these details to your iPad through the iTunes software. If you have not already installed iTunes, download it from Apple's download website, making sure that your iPad isn't associated with your personal computer when you install the program.
- If iTunes is currently installed on your pc, connect your iPad through the provided USB cable and go through the Info Tabs near the top of the summary web page.

- Given that you're at the information tab, you will notice a tick box to Synchronize Your Contacts. Make sure that iTunes is defined to synchronize with your windows contacts; tick the package and press synchronize at the bottom left part of iTunes.
- Your contacts will now be shown on your iPad! If you have signed directly into iCloud on your mobile phone, your contacts will now start copying to Apple's cloud storage space service.

How to Import Contacts from a Windows Phone

To transfer contacts from a windows phone to your iPad, you'd first move the contacts to your home windows contacts program on your laptop. That is a reasonably reliable method, and also to learn more, keep reading!

- If you're migrating from a home windows phone and your mobile phone is linked to the internet, you ought to have Home windows Live (or Hotmail) accounts already created on the smartphone. Under these events, your contacts can be stored on the Home windows live website by default. Go through the link and register with your home windows e-mail address and security password when prompted.
- Once you can see your contacts list, click a button near the top of the web page and choose Export from the dropdown list. Your contacts will begin installing as a .csv file on your personal computer.
- On your PC, go through the start menu and open up contacts.
- From the very best of the connection's windows, press import and choose CSV as your selected file

type. Press import to save.

- Go through the search button to get the downloaded duplicate of your home windows live contacts. Once you've located it, press Next to start importing the contacts to your laptop's address folder.
- Given that your contacts are kept in windows contacts, you would be able to synchronize this data to your iPad through the iTunes program. If you haven't already downloaded iTunes from Apple's download web page, please do.
- If iTunes is currently established on your pc, connect your iPad via the provided USB cable and go through the ***INFO*** tabs near the top of the summary web page.
- Given that you're at the info tab, you will notice a

tick-box to sync your contacts. Affirm that iTunes is defined to synchronize with your Home windows contacts, tick the field and press ***SYNCHRONIZE*** in the bottom left part of iTunes.

- Your contacts will now show up on your iPad! If you've signed on directly into iCloud on your mobile phone, your contacts will now start backing up just as much as Apple's cloud storage space service is enabled.

How to Add Contact to Your iPad Manually

We have discussed uploading your contacts from your previous device; however, when you begin using your device, you will want to add contacts as you go and edit or update the info of individuals you already have. Don't worry; you will become familiar with that now.

How to Add a New Contact

To include a completely new contact on your iPad, follow the instructions described below:

- Tap on the Contact App on your home Display.
- You might see any previously existing contacts on your display. To include a brand-new contact, select the blue+ at the very top right-hand nook.
- Enter the name of your brand-new contact in the areas supplied near the top of the screen. To add a mobile number, tap **add mobile**. Tap where it says Telephone to input the number, and you'll change the label home to a choice of yours by tapping it and selecting your desired from a list. To include an electronic email address, tap add E-mail, so that as you scroll down, you might see areas for

additional input information, comprising home address, birthday, or even established custom ringtones and message shades for the contact.

- If you are satisfied with the info you have in your brand-new contact, tap ***completed*** at the very top right-hand nook to save the contact.
- Tap All Contacts at the very top left-hand nook to go again to your contact list.

Once you have stored your contact, select + to feature or add every other or tap the home button to come back to your home screen.

How to Edit iPad Contact

Editing a contact on your iPad isn't expected to vary from including a new one, just can be seen barely in yet another way.

To edit a contact:

- Open up the Contacts application from the home display.
- Please scroll down and select the contact you want to edit, to open up it.
- At the very top right-hand corner, tap ***Edit***.
- Now you can edit the contact's details as explained above, adding or changing the info as required. If you want to delete any data from a contact, select the pink group icon left of the sphere and tap delete at the right of the range.

If you wish to delete the contact completely, scroll downwards and select ***delete the contact***.

CHAPTER 3

Wi-Fi & Mobile Networks Set-up

Would you like to connect your iPad to the internet before you begin the utilization of several features, like email and the application store? Right here is a way for connecting your iPad to a guaranteed wireless network as well as your mobile data network for access to the internet.

You might have recently been linked to Wi-Fi through the preliminary iPad setup, however, if you didn't, or want to get on a particular wireless network, then this section of the manual is the correct one for you!

Enabling and Disabling the iPad Internet Connection

How to Turn OFF your Wireless Connection

If your wireless connection is slow, you may want to turn it off for a short time to let you use mobile data as a substitute - be aware of lots of information you are probable to apply! Apple has made it very quick and smooth to do this.

Open Control Centre by swiping up from below the Display Screen. In Control Centre you will see a row of six spherical icons which might be White when the function is turned **ON** and Gray while it's **OFF**. The Wi-Fi image must be the second icon from the left, so tap this to put it out.

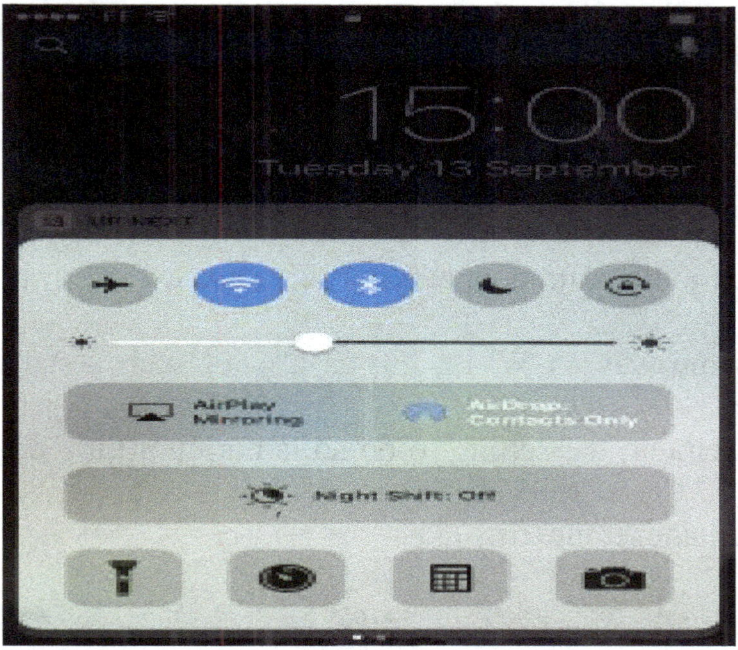

As soon as you have completed that, the wireless symbol next on your network name at the top of the display will disappear, displaying you're no longer linked to the internet. In case your iPad has a Sim card in it, the wireless image will be replaced by the data connection indicator (4G, 3G, E, GPRS) and you will be back again on Mobile Network. Just do not forget to turn your Wireless ON back to keep away from those pesky data charges!

How to Turn OFF Mobile Data

Your mobile data connection can be turned off in much the same way as Wi-Fi, but as it's not such a standard requirement, the setting to do so is buried a little deeper in the handset menus.

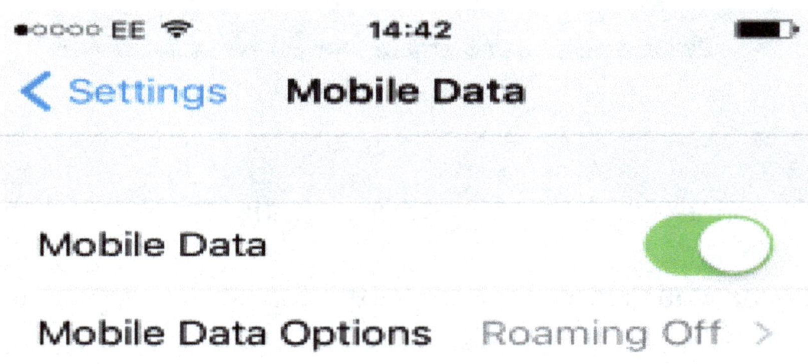

1. From the Home screen, locate and tap on Settings.
2. In the top section of the main settings menu, select on Mobile or Cellular (you would see one word or the other, depending on your mobile network).

3. At the top of the Mobile Data menu, there is a switch for Mobile Data. When enabled, the switch shows green. To turn off your data connection, tap the change.

With your data connection turned off, your phone will only access the internet if you connect to a Wi-Fi network.

Connecting Your iPad to a Wi-Fi network

For connecting your iPad to a Wi-Fi network, you'll first need to find the security key for the network. This may be on the sticker at the back or source of your router, and it might be called a WEP Key, WPA Key, or Wi-Fi password. If you're uncertain, you could check up on the person that installs your network, or your web service provider.

When you have this data, you are equipped to start!

- From the home display screen, tap on the configuration's icon.
- Within the settings menu, select ***Wi-Fi***.
- Make sure the WiFi switch is preparing to ***ON*** (green) if it's not from inception, tap the change to ***enable/allow*** it. Using the WiFi ***ON***, your iPad will check out and screen all available systems. Choose your network's name from the list shown and tap on it.
- When prompted, enter the ***Wi-Fi*** security password. That is delicate, so be sure you don't mistype it, so when you are ready to continue, tap **Join**.

- When the iPad is installed to the network, you might visit a blue tick shown up on the network's name, and a radio image will be observed next to your mobile network's name at the very top level of the screen. Whenever your iPad is at the range of the network, and wireless is switched **ON**, it'll connect automatically.

How to Connect Your iPad to Mobile Data

If you procure your iPad on a promo, you might in all probability have a month data bundle incorporated with the agreement. Allowing you to apply the internet if you are far-away from any Wi-Fi systems. This is set up automatically when you initially start your iPad, and that means you should manage to connect anywhere as long as there's a stable mobile network transmission strength! If this is not the situation, follow my brief steps below to discover the best way to get the mobile internet ready for use.

- From the home display, Tap on the configuration icon.
- In the predominant configurations listing, tap on Mobile data (depending on your network, you might see Cellular data as an alternative).
- Ensure the *Mobile data* is defined **ON** (green), tapping the change to allow it if required.
- If this hasn't worked well, you might enter configurations manually for your unique network operator. To get into these configurations, scroll right down to Mobile data network and tap on it to gain access to an option generally called ***APN***.
- You could additionally have the ability to re-download the configurations from your Sim card

by scrolling to the low area of the APN configurations website and tap reset configurations.

- If none of the strategies gets you connected to a mobile data network, I would suggest contacting your mobile network issuer for additional help, as they are capacitated to sending the configurations without delay to your device from their end.

I'd suggest most effectively the utilization of Mobile internet for email messages and general web surfing. If you watch many movies or pay attention to many online pieces of music, you might use your computer data bundle very quickly and turn into getting billed extra sums in addition to your month-to-month invoice. Test with your network service provider to discover your computer data charge, and look at the telephone bill app to know your recent data for each month.

CHAPTER 4

How To Use App Store To Find New Apps

Do you want to download apps to your iPad? With such a lot to choose from in the app store, it's hard to realize where to begin! Right here's the way to get the tasty morsels of application from the Apple application store for your iPad.

How to Download New App on iPad

Downloading new applications via the app store is a reasonably direct process; however, when you download your first app, you can have a bit of set up to do in regards to your **Apple ID**. Comply with the instructions underneath, and you will have your iPad ever downloading apps right away!

- To begin, tap the App Store icon at the home

screen of your new iPad.

- The store will open up with the Featured Apps web page. This may display to you the applications presently being promoted, either by Apple themselves or via the App developers. You can scroll down the page to look at distinct sections, and each areas' icons can be swiped through to examine which apps are being featured.
- Tapping **See All** at the right-hand side of every featured segment will show you that option in more detail.
- At the top left of the first website, you could select on **Categories**, which breaks up the App Store into broadly titled segments, without difficulty navigating through the sections. Tap **Cancel** at the top to return to the presented apps web page.
- In the categories, you may see sub-categories to

make it simpler to browse the kind of apps you're looking for.

- At the bottom of the leading web page, you'll find links to various sections of the App Store. Featured you will see, and top Charts which are self-explanatory - brief access to **"Top 40 applications"** style lists of free and paid applications if you want to scroll through.

Explore allows you to look for what human nearby are downloading, which can be quite useful when trying to find your way around a new location, or in case you're at a sporting or musical occasion.

Tapping **SEARCH** at the lower part of the display screen allows you to enter the name of an App you have heard about or recommended to you.

Updates, as the name implies, is where you can control your apps and download updates.

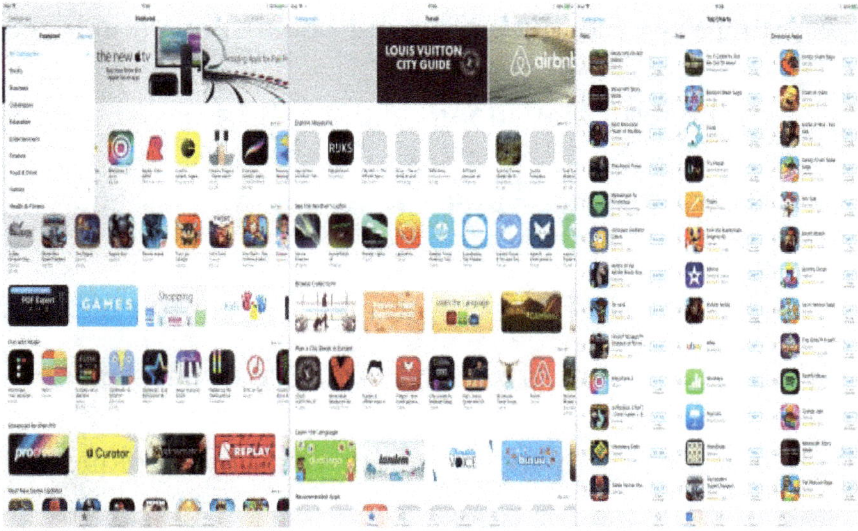

- When using the search function, anything you type into the field causes results to auto-fill on the web page. Tap the Search suggestion you like, to see what apps it brings up.
- If the button to the right-hand side of the app's name says **Download**, the app is free, and there will be no price to yourself. Tapping the word **Download** will change the box to say **Install**.
- Tapping Install will start the download process. If

there is a fee, the price would be displayed in place of the term **Download**, and you'll be asked to link a card to your **Apple ID** to pay for this and any future purchases.

- If that is the first app you have attempted to download on your new iPad, you may be requested to input your **Apple ID** information. If you have already got an **Apple ID** (if this is your first iPad you will be prompted to create one at some stage in the initial set up process), you can tap use existing **Apple ID**, and **Sign in**. In case you do not have an Apple ID, select create a new Apple ID and observe the on-screen instructions to create your free Apple ID account. When you have already introduced your Apple ID by signing into iCloud, you will need to put in your password.
- For ease of use, you may set a time hold of 15 or

more minutes before your password is needed again. This can make it smooth to install several apps in one session as you may not need to enter your password every time. Tap to regularly require your security password or require after 15 minutes, as you deem fit to you.

- You'll necessarily need to accept the iTunes Terms & Conditions, and the Apple privacy policy (which you can get dispatched to yourself via email by tapping the supplied hyperlink) by tapping **Agree**, then show that you genuinely do agree by tapping **Agree** again!
- With your account signed in and the agreements handled, your app can now be downloaded. Tap ok for this to happen.
- When the application is downloading, you'll see a blue development indicator in the shape of a circle,

with a square at the middle. Tap this if you want to pause/restart the download for any cause.

- When the application has been successfully installed, the progress button changes to an open button to which will let you access the app. When installing, the app will also appear in the first available home display area. However, you can without difficulty move this in case you want!

While you've completed your task in the store, press the **Home button** at the bottom of the iPad's front panel to return to your home display screen. Swipe across to see your newly installed apps!

How to Manage Apps on iPad

How to re-arrange iPad Apps

Whenever you download a brand-new app, it is automatically going to primarily occupy the next available space on your home display. You can easily re-arrange the applications into any order you want. To try this, tap and hold your finger on the application icon for some seconds. All the icons start jiggling. Now, all you want to do is place your finger on the image you need to move and drag it to the precise position of your choice. You can walk in between the display screen by moving the icon at the edge of the screen for a 2nd or 3rd. When you have finished, press the home button to go back to the standard display.

How to Organize iPad Folders

Setting your apps in folders makes it loads quickly to search out the app you're searching. In preference to scrolling through pages of apps, you could click on the appropriate folder and go immediately to the app you want.

To create a folder, all you need to do is place your finger on an app icon until it starts to jiggle similarly to when you are re-arranging apps. Then pull and drop an application icon at the top of another app icon. This will place both applications in a folder. You may change the name of the folder and move in a few more apps if you need. If you have completed the process, press the home button.

To move objects out of a folder, open up the folder first

then maintain your finger on one of the application icons in it until they jiggle, then tap and hold on your selected app, and drag it out of the folder to the home screen. If you pull the remaining last app out of the folder, the folder itself will vanish.

How to Delete iPad Apps

When you have downloaded an application which you do not like or that you don't need again, you may delete it off your iPad. To do that, just press and keep your finger on the app icon till the icons begin jiggling about. You ought to then see an **X** at the top left corner of every icon.

Tap the **"x"** to dispose of the app. Do not worry; you cannot take away inbuilt apps on phone or contacts by doing this so that it won’t cause any problem.

CHAPTER 5

How to Secure iPad with Lock Screen

On an iPad, you have a preference among a Custom Alphanumeric Code (that is a password with the use of letters and numbers), a Custom Numeric Code (figures only, however as many digits as you prefer!) or a 4-Digit Numeric Code (a fantastic old style pin!). You will need to decide which you need to use, so it is worth considering that earlier than you dive into the settings.

- Tap on the configuration icon, then scroll down and Tap on ***Touch ID & Password***.
- In the **Touch ID** & Password menu, tap the blue hyperlink to turn Password ON.
- The default Security password placing is a Custom Alphanumeric Code - a complex password

containing letters and digits. You could alternate this by tapping Password options.

- Tap your chosen password option to select it.
- Enter your **PASSWORD**. While you type in your secret four-digit, the display screen will increase automatically.
- Re-input your Security password to verify it. If the entered Passwords do not match, you might be returned to the first Security password access display to start over. If the Password that has been entered matches, then you will go back to the Security password menu.
- The final element to decide is how fast you want to enter your Password, which is often a balance between usability and safety. To change this setting, Tap **Require Password**.
- Pick your time out from the listing on screen by

tapping the interval you want to set. A tick will appear on that line, and when you're happy with the setting, tap Back at the very top left-hand nook.

You can allow access to certain functions of your iPad when the screen is locked. There are switches to enable the usage of ***NOTIFICATIONS, SIRI***, and other components of the working gadget. Tap any of these to permit them (when **ON**, the switches can look green).

How to Set Up Touch ID to Unlock Your iPad

Now that you have set up a Password, you may want to enable **Touch ID**, which is fingerprint recognition to unlock your Devices (meaning you might not need to type in that password, even though you could if you want to!).

Follow the instructions below to achieve this effortlessly;

- To begin setting up **Touch ID**, you will want to be within the Settings **Menu**.
- In the predominant settings menu, select **Touch ID & Password.**
- Input your **Password** to access the settings.
- Tap **Add a Fingerprint**.
- To start including your fingerprint, place your finger or thumb onto the ***Home button***, however, do not press it. Lift and replace your finger as instructed on-screen, shifting it very slightly as you achieve this. When the center of your fingerprint has been scanned, you'll be requested to place your finger in unique positions on the home button to experiment the edges.
- While your fingerprint is fully scanned, you will

see the whole screen, displaying that your print has been captured and Touch ID is ready. Touch **Continue**.

- As you add fingerprints, they'll be numbered. You can change the names (so you recognize which print is which) or delete fingerprints from the phone by Tapping the name after which editing or deleting as required.
- When you've modified the name of fingerprint, Tap **Done** on the keyboard to keep the new name.
- **TOUCH ID** will be without delay activated for unlocking your iPad, and for **Apple Pay**. To enable iTunes and App Store Purchases to be authorized alongside with your fingerprint, Tap the switch to carefully turn it **ON,** then get into your Apple ID password.

You can upload as much as five Fingerprints, so putting

in place fingerprint access for your family members may be carried out too. A phrase of warning though; remember that in case you've introduced the capacity to use fingerprint scanning to authorize iTunes and App Store buys, anybody who's fingerprint is added can do this too!

CHAPTER 6

Sending Emails & Attachments from iPad

To send an email, you ought to have already created an email account on your iPad.

- Find and open the app on your home display screen. This looks like a White Envelope, and if you've acquired emails already, there may be a pink badge on it which represents some unread email messages.
- The email App will open up your brand-new E-mail. To view your inbox, Tap **INBOX or ALL INBOXES** at the top left-hand nook of the screen.
- When you can see your E-mail inbox, select the **COMPOSE BUTTON**. This looks like a pen and

paper, and it's located at the top-right nook of the email inbox screen.

- The screen will now show blank email, ready to begin writing.
- New emails will regularly send from your default electronic mail account (which is usually the first one you have added). When you have multiple accounts on your iPad and want to switch the accounts to send from, it is easy to do. Touch the **CC/BCC, FROM** collection which shows the e-mail address with which you're sending from, then Touch the e-mail address shown to change it to another account.
- To add a **recipient**, Tap into the **TO** field. To browse your contact list, Tap the + button at the right-hand aspect of the display. You could additionally start to type their name, and any

matching contacts would be shown allowing you to pick the one you are searching out for. In case you don't have the Tap saved for your device already, simply Tap the **TO** Field and start typing the E-mail address you wish to send to.

- Utilize the **SUBJECT** box to add a title to your e-mail.
- Tap into the main window (above the pre-loaded send from my iPad signature) to put the cursor there and type your message. When you're ready to send your e-mail, select **Send** at the top right-hand nook.

How to Add an Attachment to E-Mail

iPads can feature photos and videos from the device as an attachment ever since iOS 7 was launched in 2013. However, you can now additionally add attachments from online storage including Google Drive or Dropbox, especially with the recent **iOS 13, iOS 12 and iOS versions** to come.

How to Attach an Image or Video From Your iPad

- Begin by creating and accessing your email account as stated above.
- To feature/attach a picture or video, Tap the **CAMERA** icon to the right of the keyboard's top level.
- Please navigate through the image folders you've created for your iPad to discover the image or video you want to transfer and Tap it to choose it.
- To choose the image or video and connect it for your email, Tap **USE**. You can attach one at a time. However, nothing is stopping you from adding any

other one!

- Provided that you've launched your internet connection, Tap **SEND** to get it sent immediately.

How to Add a Document From an Online Storage App

To add accessories from a web storage account such as **Dropbox or Google Drive**, you'll need to have the App for that service installed on your iPad and be signed directly into your account in the App to gain access to your documents.

- Create your email as described above, then select the paper clip icon to the right of the keyboard's top line.
- By default, your **iCloud Drive Storage** will open. If that is where your file is, navigate through your folders and choose it. To import an item from external storage, Tap **SOURCE**.

- The first time you use this feature, you'll want to allow access to which you will use an option storage space source. Tap **More**.
- The manage location display allows you to select the storage Apps you want to include documents from elsewhere. Tap the switch to the right of your preferred App to allow/enable it (it turns green) then Tap **DONE**.
- You'll then be back at the **iCloud Drive Screen**. Tap Source once more to open up your storage choices.
- This time all the storage space Apps you enabled in the previous display screen should be accessible for you in the menu. Tap the Appropriate one to discover the files and folders therein.
- Please navigate through the folders to locate the file you are considering, and Tap it to attach it to your email.

- When you've attached your document(s), Tap **SEND** to get the document attached and sent to your contact(s)!

Chapter 7

How to Create a Folder on an iPad

We've all been there (searching through web page after web page of application icons looking for where we put our Facebook application or that favorite game we haven't used in some time). The best thing about the iPad is just how many amazing applications you can download, but this includes a price: a great deal of applications on your iPad! Fortunately, there's one great technique for maintaining your iPad structure (you can make a folder for your apps).

Developing a folder on the iPad is one particular task that truly is as easy as 1-2-3.

- *Grab the application with your finger*: If you aren't acquainted with moving apps around the iPad display, you can "grab" an application by keeping your finger onto it for a couple of seconds. The application icon will increase somewhat, and wherever you move your finger, the application follows if you keep the finger down on the display screen. If you wish to move in one display of apps to some other display screen, simply move your finger to the edge of the iPad's screen and await the display to change.

- *Drop the application on another application icon*: You develop a folder by dragging an application onto another application you want in the same folder. Once you grab the app, you make a folder by dragging it together with another application you want in the same folder. When you hover together with the destination app, the application would blink once or twice and then broaden into a folder view. Drop the application within that new folder display screen to produce the folder.

- Name the folder: This is the third step that's not too necessary. The iPad gives the folder a default name like 'Video games,' 'Business' or 'Entertainment' when you create it, but if you would like a custom name for the folder, it is simple enough to edit. *First, you need to be in the folder view; you can leave a folder by clicking the Home button. On the home display, keep your finger on the folder until all the applications on display are jiggling. Next, lift your finger and then click the folder to increase it. The folder name near the top of the display screen can be edited by tapping into it*. After you have edited the name, click on the Home button to leave the 'edit' mode. You can add new applications to the folder using the very same method. Just grab the application and move it together with the folder. The folder would broaden just like it did when you initially created it, letting you drop the application anywhere inside the folder.

How to Remove an App From a Folder or Delete the Folder

You can remove an application from a folder by merely doing the reverse of what you did to generate the folder. You can also remove an application in one folder and drop it in another or even create a new folder from it.

Grab the app: You can grab and move applications around within a folder just as though the applications were on the home screen.

Drag the application from the folder: In folder view,

there's a curved box at the centre of the display that signifies the folder. If you pull the application icon out of the package, the folder will go away, and you'll be back again on the home display screen where you can drop the application icon anywhere you desire. This includes shedding it into another folder or hovering over another application to make a new folder.

The folder is deleted when the last application is deleted: So, if you would like to delete a folder, move every one of the apps from it and place them on the home display or in other folders.

Organize Your iPad Folder as You want

The best thing about folders is that, in lots of ways, they act exactly like app icons. This implies you can pull them in one screen to another or even move these to the dock. One cool way of arranging your iPad is to separate your applications into different categories, each using its folder, and you'll be able to move each one of these folders to your dock. This enables you to have indeed a single Home display screen that has usage of all your apps, or you can merely create one folder, name it 'Favorites' and then put your most used applications in it. After that, you can place this folder either on the original Home display or on your iPad's dock.

Chapter 8

How to Locate Saved Files on an iPad

Your iPad is like having a slim-line PC with you all the time; it could store all of your favourite applications and files, which makes it easier to access essential documents wherever you are in the world. You should understand and know where to find downloads on your iPad and how to control your documents.

Where Are Downloads on My iPad?

Locating where the download is on your iPad is quite easy when you are using iOS on any Apple device, but that isn't the case as it pertains to managing data files. There is no iPad downloads folder where all documents immediately go to like on Personal computer or Mac, and the iOS document system isn't as effortless to search like a Google android file system.

In which a file downloaded is reliant on the app you utilise to see it with, although things have grown to be just a little simpler to understand because of the introduction of the Files application in iOS 11.

Choosing where you can Save a File

It is critical to save your data files to a spot where you can find it again. However, there are various alternatives for saving documents, but here's how to save lots of documents easily from popular apps.

Open up the relevant email: Faucet Options in the very best right of the display, highlighting your options dialogue box

Choose which application to send the document to With regards to the document, you can usually faucet Save Image for images, duplicate to iBooks for PDFs or Save to Documents to save lots of it to the Data files application for general use. The iPad Email app shows email and attachment, highlighting the Save to Documents option. *Tap the greater icon to find more options.*

If you tap Save to Files, after that you can choose to either save the document to your iCloud Drive or right to the iPad, then tap Add; iPad Email App displaying where you can save a connection to if you wish to access the document from other iOS or Mac devices, touch iCloud Drive. Done well, you've successfully preserved the document to your selected location!

Saving a Document from Safari

Here's how to save lots of a document from the default browser, Safari.

- *Open the document in Safari*
- *Tap Options:* iPad Safari application showing where the Options dialogue is on the browser

Choose where you can save it: The Safari browser running with an iPad, with the Save to Data files dialogue box highlighted

You may want to scroll to find more options such as *Save to Files*, depending on how many apps can be found to use the file.

Saving a graphic From Safari

Open up the image in Safari: Keep your finger to the image, then release after an instant or two to talk about the dialogue package.

Touch Save Image to save lots of the image to your Photos folder: A graphic within Safari on the iPad, with Save Image highlighted.

Where to find Downloads on your iPad

Once you have downloaded a file, where do you think it is? Here are the many places your file might be:

Images: if it's an image file, it's probably stored inside your Photos Gallery.

PDF: saving your “file to iBooks,” PDF documents are sent or copied to iBooks.

Other Documents: For all the data files, it's probably stored in the Data files App. This application brings together all of your documents across iCloud so that it can likewise incorporate documents from your Mac or other iOS devices.

Chapter 9

How to Get More Things Done on an iPad

Perhaps you have ever wondered if there is an instant way to get more things done or a much better way to achieve certain things on the iPad? Every year, Apple produces a new version of the iOS operating system that works on the iPad, and with each new version, features launched can increase efficiency by assisting you to do specific jobs faster and better. There's only one problem: not everyone knows about these new features. We'll review some of the secrets that showed up with the initial iPad, including some which have been added over time to help you navigate the iPad just like a pro.

- Tap the Name Bar

We'll take up a critical suggestion that will help increase your ability to control your iPad. Perhaps you have ever scrolled down an extended list or been in the bottom of a big website and needed to go back to the top of the website page; there is no need to scroll. More often than not, you can select the title tab of the application or website to come back to the start of the list; this works together with most apps & most webpages, although don't assume all web page is created to be iPad-friendly.

- Manage Music group, iMovie and iWork

Do you realize a whole collection of secret applications include the iPad? Going back a couple of years, Apple has made the iWork and iLife collection of applications free for individuals who buy a new iPad. These applications include:

- *A word-processing app.*

- *Figures, a spreadsheet.*

- *Keynote, an application for providing presentations.*

- *A music studio room with virtual devices.*

- *iMovie, video-editing software that includes some fun templates.*

...and many more.

- Download Free Books on your iPad

Everyone loves free stuff! And you may get lots of freebies with your iPad if you know and understand where you can get them. For book enthusiasts, the best-held key on the iPad originates from something called *Project Gutenberg*. The purpose of Project Gutenberg is to consider the world's library of general public domain works and convert these to digital. *Treasure Island,*

Dracula, Alice in Wonderland, and Peter Skillet are only several books you can download free of charge on your iPad.

If you haven't already done so, you'll first need to download the iBooks software. That is Apple's digital bookstore and audience. After you release the iBooks app, tap the very best Charts button in the bottom of the screen which will provide two lists: the very best Paid books and the very best Free books.

Click on the Categories button at the top-left part of the display screen; this will drop down a categories list. If you're thinking about reading a few of the most excellent books ever without paying a dime, choose Fiction or Non-Fiction and Books from the list. You may now have the ability to scroll through typically the most popular free novels available through iBooks. You can further thin the list by choosing a category like Sci-Fi & Dream or Young Adult.

- Let Your iPad Read Chosen Text for you

Would you like to give your eye an escape and let your iPad do the massive raising? The iPad can speak selected text messages for you, but first, you'll need to carefully turn on this feature in the configurations. The text-to-speech feature was created to help the eyesight impaired, but it could be beneficial to many people. For instance, the iPad could enable you to multitask by reading a fascinating news article for you while you're preparing dinner.

How to Start the iPad's Text-to-Speech Feature

1. First, go into the iPad's settings.

2. Next, choose General configurations from the left-side menu.

3. Tap Convenience from within the overall settings. It is right above the section for Multitasking Gestures.

4. From within the Availability settings, choose Conversation. This is the latter in the Eyesight block.

5. Start Speak Selection by tapping the associated slider. This establishing will put in a new Speak option to the menu that shows up when you decide

on the text.

6. If you believe you might use the feature often, you can also start Speak Screen; this enables you to slip two fingertips down from the very best of the screen to read the whole display to you; this won't work so well with webpages where the surface of the display is filled up with selections, but it is useful with other applications like Mail.

7. You can even change the voice utilized by tapping the Voices button within the Speech settings. Also, focus on the Speaking Rate. This is adjusted to have the iPad speak faster or slower.

One smart way to use the text-to-speech feature is at iBooks, where the iPad can browse the reserve to you. This is not quite as effective as a publication on tape, where the reader can provide the right inflection to what or even portray the character's voices. However, if you opt to speak the display screen, the iPad will automatically change pages and continue reading the book.

Chapter 10

How to Use Drag and Drop Features on an iPad

The drag and drop feature involves a multitasking process, and the frequent need to use multiple fingers or both of your hands on the iPad at the same time to move files. Drag and drop is an option to the popularly known *copy-and-paste*; when you move a file from one directory to some other folder on your computer, you are just performing a copy and a paste by making use of your mouse rather than menu instructions; and with the iPad already assisting a standard clipboard, you can duplicate an image from the Photos application to the clipboard, open up the Notes application and paste it into one of your records.

Why do we need Drag and Drop feature?

First, drag-and-drop makes the procedure smoother when you're able to start the Photos application and the Notes application side-by-side and move photos from one to the other. Moreover, you can grab multiple pictures and pull them all simultaneously to the destination folder. This makes it possible to choose multiple images and submit an email reasonably simple.

You can also choose photos from multiple sources, and that means you can grab an image in the Photos app, open up Safari to include an image from a website and then open up your Mail application to drop them into a note.

What to Drag and Drop on an iPad

So, what are the possible items you can pick up?

You can pick up just about anything that may be thought

of as an *object*. This consists of pictures, documents, or even chosen text. You can even grab links in the Safari internet browser and drop them into a text, an email, etc. You can also grab a text document from iCloud Drive and drop it into Notepad where it'll show up as the material of the written text file.

Move and drop works both within the same applications and across multiple apps; for instance, you can get a web link in Safari while in scenery setting, move it aside of the display and drop it in to the space that is established to start a break up view of both websites in the web browser, or you move the same hyperlink into a fresh message in the Email app.

How to Drag and Drop on the iPad

The actual notion of drag-and-drop is straightforward,

but its implementation happens to be (and could remain) complex. Dragging an object just like a document or a picture from one place to another is as easy as moving your finger, however when you consider multiple items and multiple apps, you may want to place the iPad on the desk or your lap and use both of the hands.

To get an object, press your finger against the display screen for minutes; the item will pop up from where it is situated in the app, once it pops out of its original place, you can move your finger around the display, and the picture or object will stay trapped to your finger. To get additional purposes, tap them with one of your other fingers; that's where using two hands makes the procedure much more workable. If you're selecting multiple images in the Photos app, you can touch each picture to include these to the items being dragged.

If you are dragging images, website link, text message or

other items, the procedure of the iPad remains exactly like if you weren't dragging anything; this implies that you can close from the application by clicking the *Home Button*, open up another application and drop your selection involved in it.

Moreover, you can start a whole new app and increase your chosen items, but you'll need to employ a new finger to get items from a different app; and, yes, you may use multiple fingertips to drag items, you can as well change fingers; if you will pardon the pun, this is a convenient feature. You can switch the selected item to some other finger on a single hand or even to your other hands; you need to place the 'new' finger near to the finger 'holding' the selected item, and you'll start to see the selection move under the new finger. When the selected item sticks to the new finger, you can lift the initial from the display screen.

As regards the dock for opening applications or multitasking, if you move a finger from underneath the iPad toward the very best, the application dock will be exposed. You should use the dock to open up a new app, or you can pull an application from the dock to the center of the display to open it up together with the application as a floating column; this will help you to use both applications at precisely the same time.

If you know you are dragging in one application and dropping into another app, it is easiest to open up both applications on the display screen first in slide-over or split-view and then move from one application to the other.

How to Use Drag-And-Drop to Transfer images for an iPad

There are a variety of great ways to use the new drag-

and-drop feature from selecting photos relating to a document or email message to grabbing selections of text from a website to drop into Notes, but possibly the most versatile is how it can connect to the Files app.

An excellent example is importing photos from your personal computer to your iPad; drag-and-drop can make this a more natural process.

Place your photos within an iCloud folder, open up Documents and Photos in split-view on your iPad, and then use drag-and-drop to move multiple pictures at the same time from the folder in iCloud to whichever album you want to put them into within the Photos app.

There is no need to plug your iPad into your personal computer, use iTunes, or even to transfer from a cloud storage space by conserving each picture to your camera or use a third-party app. In the recent iOS version, it is only a simple drag-and-drop process.

The capability to copy files and photos so easily can be incredibly useful after the Files app facilitates third-party cloud storage services like Dropbox, Google Drive, etc.

CHAPTER 11

How to Proceed When You Can't Activate Used iPad

If you buy a used iPad, it is interesting. In the end, you come with an iPad and stretch your budget by acquiring a used one, especially for individuals who are not economically buoyant.

Some individuals encounter this issue along the way of activating their new device: The iPad will inquire further for somebody else's Apple ID and wouldn't typically work unless supplied.

This isn't a challenge that can’t be fixed, so do not fret because you'll get it fixed following these steps.

- It is consequently an attribute of Apple's Find my iPad service known as activation lock.

- Activation Lock is a security measure that Apple raised to cope with the allergy of iPad thefts. In earlier years, if someone takes an iPad without blockage by lock feature, they could clean it, resell it, and breakout with the crime. Activation lock altered the situation.
- When the initial owner setup finds my iPad on the tool, the **Apple ID** used will be stored on Apple's activation servers together with almost every other information about the phone. The activation servers will most effectively unlock the phone again if that unique Apple ID can be used. If you no more have the Apple ID, you'll never be in a position to activate or use the tablet. This facilitates the security of your iPad because nobody would like to grab a tablet they can’t use. On the other hand, it generally does not harm you

if you recently procure the phone.

- Dealing with activation lock is annoying, but additionally, it is smooth to solve. It's mainly possible, and the prior consumer just forgot to carefully turn off find my iPad or erase the tool correctly before offering it on the market (though it could also be a sign you've purchased a stolen device, so be cautious).
- You should contact the preceding owner of the telephone for him/her to consider the necessary steps.

How to Remove Activation Lock on iPad

- It is expedient that you should unlock or remove activation lock from the acquired iPad (used iPad) by inputting the prior owners' **Apple ID.** This technique can be initiated by getting in contact with the owner and detailing the scenario.

- If the owner lives near to you, I'll recommend that you hand over the phone back to him/her with the mission to insert the mandatory unlock code which is his/her Apple ID. When the seller gets the iPad at hand, he/she only will enter the necessary Apple ID on the activation lock display. Having done such, restart the telephone and then forge forward with the typical activation process.

Ways to Remove Activation Lock using iCloud

Sometimes, things can get a bit messy and complicated if the merchant/seller cannot physically access the tablet thanks to circumstances such as distance among other factors. This may also be resolved effortlessly as the owner may use iCloud to eliminate the activation lock from the phone through his accounts by following the steps below:

- Visit iCloud.com on any device, either mobile or laptop.
- Log-on with the Apple ID he/she used to activate the telephone.
- Click Find My iPad.
- Select All Devices.
- Go through the iPad you sold or want to market.
- **Select Remove from Accounts.**

Having achieved that, after that, you can PULL THE PLUG ON the iPad, and you switch it ON again. After that, you can proceed with the standard activation process.

How to Fix Locked Home-Screen or Security Password

If you activate your phone and find out either the iPad's home display screen or the security password lock display, therefore that the supplier/vendor didn't completely erase the smartphone before offering it for you. On this notice, you'll need the owner to wipe these devices to be able to do it with the activation process.

The next two procedures should be followed as you hand over the phone to the owner or seller to unlock the phone;

- If the tablet works on iOS 10 and later version, the owner has to log out of iCloud and subsequently erase these devices by heading *to Settings -> General -> reset -> Erase All Content* and *Settings.*
- If the tablet works on iOS 9, the seller/seller must go to *Settings -> General -> reset -> Erase All*

Content and *Settings* and enter his/her Apple ID when prompted.

- When the erase process is completed, you're absolving to activate your phone with no further ado or hold off.

How to Wipe an iPad Using iCloud

Imagine if you can't gain access to the vendor/merchant due to some reasons, yet you will need your mobile phone to be wiped entirely for easy convenience, the seller may use iCloud to erase it. This is attained by ensuring the phone you want to get triggered linked to a WiFi network or mobile data network, and then inform the seller to follow along with the next steps:

- Visit [*http://iCloud.com/#find*](http://iCloud.com/#find)

- Sign in with the Apple id he/she applied to the phone that is with you or sold to you.
- Click *All Devices*.
- Choose the phone sold you or available to you.
- Select *Erase iPad*.
- When the phone is erased, click *Remove from Accounts*.
- Restart the phone, and you are all set.

How to Erase an iPad Using Find My iPad App

This process is very much indeed identical to the approach explained above using iCloud by just using the Find my iPad application installed on some other iPad device. If the owner prefers to get this done, connect the phone you're buying to Wi-Fi or mobile data, and then inform the owner to adhere to the steps below:

- Start the *find my iPad* app.
- Sign on with the Apple ID they applied to the phone sold to you.
- Choose the phone.
- Tap *Actions*.
- Tap *Erase iPad.*
- Tap *Erase iPad* (It is the same button, however on a new display).
- Enter *Apple ID*.
- Tap *Erase*.
- Tap *Remove from Accounts*.

Restart the iPad and get started doing the setup process.

CHAPTER 12

How to View Pandora Channels Offline

Are you a **Pandora** lover? I will help you make your playlists to be accessible offline. Doing work to save lots of some music on your mobile phone doesn't take much space for storage to your device, and stored tunes is an exquisite component to have access whenever you're short of data bundle/tariff connection but need to hear music to resolve your boredom problem. The feature works on **Google android** and **iOS devices**.

If you've not made your playlists available offline, doing this is quite exceptionally easy and can be carried out in only a few moments.

One important caveat: You need to be considered a paid customer to Pandora through Pandora Plus ($5/month) or even to Pandora Premium ($10/month.) you can view the programs on Pandora's website.

Without further ado, follow the instructions below;

- I recommend linking your phone to a wireless connection. You can download songs more than a mobile data connection instead of cellular, but it's heading to have a reasonable level of data utilization to get the whole lot downloaded. When you have the option for connecting to a WiFi network, please do this because it can save you more time with regards to the truth that Wi-Fi is faster than mobile data generally in most conditions.
- Start the Pandora app.
- Making channels available offline requires you to

undoubtedly have channels to be produced offline. If you haven't made any channels on Pandora, have a short while to make a few. You'll also have to pay attention to them for at least a few tunes so that Pandora considers them your favourites.

- Tap the three lines located at the very top left facet of the app so that you can gain access to Pandora's menu. At the lowest area of the screen, you'll see an "offline setting" slider. Slide that pub to initiate offline setting for your tool. When you do, Pandora will synchronize your top 4 channels on your smartphone and lead them to be accessible offline.

At first attempt, I'd advise that you let your mobile phone stay connected to the Wi-Fi network for approximately 30 minutes to make everything synchronize. How fast things happen depends on the speed of your web connection.

When the whole tune is synchronized, on every occasion you will need to hear tunes offline, you need to visit that menu, and you will toggle the offline button ON. The application will remain in an offline setting until you situate it back to the conventional configuration.

Why Should Pandora Station be Used in Offline Setting?

You can focus on Pandora each day you are at home. You could have a radio train station for when you're running, another for enough time you are strolling out with your dogs, plus some other for if you are working from home doing some jogging. You can similarly have a route you dedicate for just about any social gathering you host.

I take benefit of offline mode because I like to visit a lot of different countries, which is an incredible enjoyment, aside from the cellular phone expenses. Whenever I am on a journey, I try to use only a small amount mobile data as you possibly would want to stay away from the high cost of use that comes by the end of the month, but slicing out specific applications like *Pandora*.

Why should I exclude using Pandora, especially in conventional mode or online mode?

That is done because streaming music occupies the right amount of data bundle, due to this, it's off-limit for individuals with limited data plans (luckily, unlimited when at home). Additionally, you lose out on hearing when you're in places like aircraft and trains where your computer data connection is either sluggish or non-existent.

CHAPTER 13

Hidden Secrets of Customizing Your Pandora Stations

When you pay attention to music on Pandora, you can experience a predicament where the options of music do not appear attractive to your interest.

You might likewise discover yourself hitting thumbs down all too often or seeking to bypass tracks. The amount of situations you can miss tunes is bound except you to have Pandora plus. You might also get bored and weary in the channels for hearing music repeatedly.

Remember that Pandora uses all the characteristics of this first seed music put into place. The monitor or designer you used to produce the train station; however, will not fit the attributes of each other music it takes on. Music is exclusive, and few tunes have the same complete features or in Pandora's conditions, the same DNA.

Possibly, Pandora is playing the music you don't like as it isn't matching the features that you want from the seed song. You might probably prefer the station; nevertheless, you want to combine it up with a few tracks with a faster tempo, or by adding country music or an oldie that may have different quality guidelines.

The Fastest Way To Change The Feeling Of The Station

After you've paid attention to one station for quite a while, you start hearing precisely the same songs. If you've become bored of your place, or if you need to hear music from your other channels, you may use **"Quick Blend."** Around the Pandora application for press streamers or network Home-Theatre device (Blu-ray Disc

Player, Smart Television, some stereo and a lot of Home Theater receivers), you can Quick Combine all your channels collectively to try out music that suits requirements from all of your channels.

Within the Pandora browser player and application smartphones, you can specify channels you want to mix to regulate the feeling of the music you're taking part in. ***NB**: Quick Blend is temporary and cannot modify or personalize these stations.*

How to Fine-tune Your Station Further by Merging Tools

If you're focused on fine-tuning your station, you need to be steady and specialized in discovering the right mix of factors to get just what you need.

- Make use of thumbs down regularly. If a track will not fit the form and desire to have that train station, Give such a thumbs down without displaying mercy. It could be challenging to provide thumbs down to a piece of music you prefer; however, which doesn't fit but be courageous; you have an objective. Thumbs down will not have negative influence on the monitor turning up on your different channels. As time passes, Pandora will dispose of characteristics you don't find appealing.
- Make use of thumbs up occasionally. This will help you to enhance the tunes that match the station.
- Ensure you create several channels. As you pay attention to your place, you may locate a piece of

music that is a fit for disposition you want to create. It would help if you used that music to make a new train station. On a press streamer or suitable compactable tool, you can select "create a place" and enter the name of the track.

Make an effort to creating a significant amount of channels using similar tracks, then apply the Thumbs down method of refining the channels. When you create the perfect train station, you can delete or take away the testing stations.

If none of these music tracks works, think of the qualities you will need within the place. Possibly music you don't love is a much better match and may create the train station.

NB: *While creating test channels, it might be expedient to group them. Similarly be sure you rename the* *channels with notice and digit to keep them alongside the place list "A01," "A02," "A03", etc.*

Ways to Get More Selection of Tunes and Mood

Conversely, it is viable to make a station with an enormous variety of songs and mood.

This may be attained by following the process below;

- Add more seed tunes or seed performers.

  It would help if you used the "add variety" button on your laptop, or you can include tracks to the train station page.

- Be generous by using thumbs up.

  The more magnificent music you prefer, the higher the features that'll be used in the decision of songs for the station, thereby creating more variety.

- Utilize "I'm Sick and tired of this Track."

  This program is of all network media players and network devices. Select this feature instead of the utilization of the thumbs down whose function is to thin down the types of music played.

The greater committed you are to the effect, the higher your potential for creating a perfect station. Customize your music because music is personal. Once you get the cling of, and funnel the advantage of Pandora's development and establishing options, you're adequately ready to manage your music listening experience.

CHAPTER 14

Steps for Fixing an iPad that won't Power ON or Charge

Sometimes whenever your iPad doesn't start, you will think you should buy a new phone. This can be true if the trouble is terrible enough; however, there are diverse methods to be applied on your iPad before you finally conclude that the phone is dead. In case your iPad doesn’t **START**, try these six steps to get the problem set yourself.

- **Charge Your Tablet**

Ensure that your iPad's battery is well charged sufficiently to power the smartphone.

If you wish to try this, plug your iPad into a wall structure charger or your personal computer via USB cable, but I strictly recommend using power. Make sure you allow it charge for at least fifteen to thirty minutes. This could make your mobile phone get start up automatically, or you can also press down the power button to carefully turn it ON.

If regardless you think your phone ran out of battery when using or recharging fails, it is feasible that your charger or cable is defective or the battery is terrible which implies you may want to acquire another battery. You can try to use various other cables to check.

- **Restart Your iPad**

If charging the battery didn't Turn your iPad ON or your iPad does not charge while ON and you're sure the charger works on other mobile phones, the next thing you have to try is to restart the phone. This is attained by pressing down the **power button** at the very top right side of the smartphone for a couple of seconds. If the smartphone is OFF, it must turn ON. If it's ON, you will notice the slider offering to turn it OFF.

- **Hard Reset the iPad**

You could attempt a hard reset if the normal restart didn't transform it ON. A hard reset is kind of a restart that clears a higher percentage of the phone's memory space (however not its storage space, you won't lose any document) for a more complete reset.

Follow the below instructions to apply hard reset:

1. Press and hold down the **Power Button** and **Home Button** at the same time. (If you work with the iPad 7 and later versions, Press down **Power** and **Volume Button** at the same time.)
2. Maintain holding down those button for at least ten

seconds (there is certainly nothing at all absurd with holding those button down for 20 or 30 secs, however, if nothing at all has occurred by holding them down, it in all probability won't work)

3. If the Shut-down slider shows up on the screen, maintain pressing the buttons
4. Until when the white Apple logo design appears, after that, you can remove your hands off those buttons and allow phone start up.
5. Restore iPad to Factory Settings.

Sometimes the best strategy is restoring your iPad to its production default settings. This technique will remove all the info and configurations on your phone (You can synchronize and backup your details), and in exchange fix a significant number of problems.

Although typically, you'll synchronize your iPad and restore it using iTunes, however, if your iPad won't activate, try these steps:

1. Plugin the iPad's **USB cable** to the connection slot, yet not into your personal computer.

2. Press down the iPad's **Home button** (while on an iPad 7 or later version, press down **Volume button**).

3. While still pressing down the home button, plug the contrary end of the USB wire into your personal computer.

4. This will result in opening **iTunes**, it'll automatically place the iPad into recovery setting, and consequently, help you completely restore the iPad.

- **Put iPad into DFU Mode**

In a few situations, your iPad might not get **Started** as it does not boot up. This may happen after jailbreaking the phone or when you try to install an iOS revise without enough space for storage or battery life.

If this is the problem you're facing, you can simply put your mobile phone in **DFU Mode** by following these processes:

1. Plug your iPad into your personal computer.

2. Please press down the **Power button** for three mere seconds, then allow it to be.

3. Press down the **Power button** and the **Home button** (with an iPad 7 or later versions, press **Volume down**) together for approximately 10 seconds.

4. Release your hands from the **Power button**, however, maintain pressing down the home button (with an iPad 7 or later versions, keep the volume down) for about 5 seconds.

5. If the screen remains dark and nothing appears noticeable, you're in **DFU Mode**. Adhere to the onscreen instructions in iTunes.

- **Reset Proximity Sensor**

There is undoubtedly another situation that can make your iPad never to turn **ON,** which is a defect in the proximity sensor that dims the iPad's display screen. This causes the screen to remain darkish even if the smartphone is **ON** rather than also near to your face.

Follow the instructions highlighted below;

1. Press down the home and **Power Button** to restart the phone.

2. When it restarts, the screen should then be working.

3. Tap the **Configurations app.**

4. Tap **General.**

5. Tap **Reset.**

6. Tap **Reset All Configurations**. This will erase all your preferences and configurations on the iPad, however, won't delete your data.

If regardless your iPad won't start after these steps; the problem is most likely too severe to revive personally. I would suggest you contact Apple support. The support staff will either repair your phone or let you know what it could cost to fix the problem.

You should check the status of your iPad's warranty before going to the repair center because that could save you additional money on the repair.

CHAPTER 15

Hot Tips & Methods to take Pleasure from more Features on iPad

Here are some guidelines to help you love the more exceptional features on your iPad.

How to Use Two Apps simultaneously with Slide Over & Break up View

Do you want to manage your calendar or any other even while you are checking your Email? No issue, this section will show you how.

Need to research something online by safari without dropping off your gained access to on iBooks or any other? Now it's easy!

Slide Over allows you to quickly use the other App without terminating the first (or departing the display) and **SPLIT UP View** will enable you to use two apps on the display screen simultaneously - forget about needing to interchange laterally!

How to Swiftly Use another App with Slide Over

In my instance, I've safari opened up on the screen, and I have to check something on the calendar:

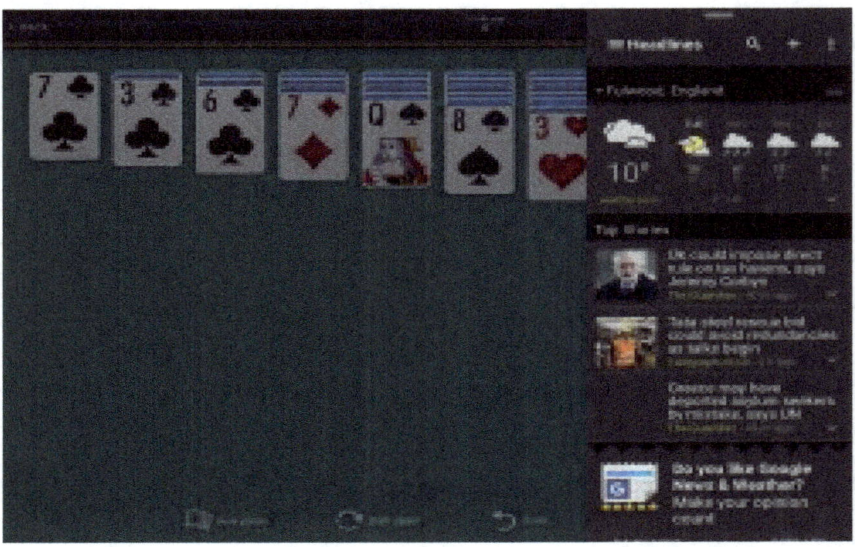

- Together with your first App open up on the screen, SWIPE in from the right-hand advantage of the display to open up **SLIDE OVER**.
- This will open up a panel running the ultimate App you used (previous in the primary windows). To alternative, this to a distinctive or new App, pull down the tiny gray pub from the very best of the Slip Over -panel. The App in the primary windowpane will dim into the background.
- Please scroll through the set of Apps to find the main one you want to open up and Tap it to open it up in slip over.
- Together with your chosen App opened up, you might now utilize it as if it's been on the original main display screen. To disregard it and go again to your earlier App, swipe the other App to the right-hand corner of the screen.

How to Use Two Apps Side-by-Side with Break up View

Having opened the next App with a slip over as instructed above, switching to a split view can be more comfortable.

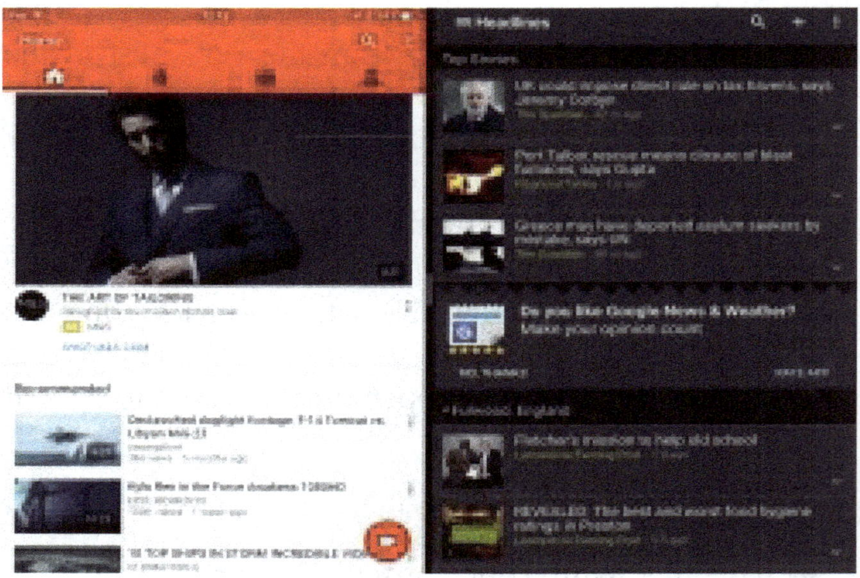

- With another App opened up in slip over, swipe the white pub toward the center of the screen. The dimmed down Background App will brighten again, which indicates that glide over has turned into the split up view. Each App is usable, impartial of each other.

- To improve the other App, swipe the grey pub down from the very best of the screen and select a distinctive App from the list. To close a divided view, pull the divider to the right-hand side of the screen.

As you observe, using split view can be a development of Slip Over, and helps it be easy to "get things achieved" without the difficulty of continually turning back and forth between two Apps. The power of the iPad method is that you will have the capacity of using high-performance Apps side-by-side without it lacking a defeat, that's brilliant if you get a Face Time Video Call if you are in the center of editing an image!

How to Manage Notifications

Notifications are your device's way of letting you know what's occurring in an App. This may be information including a note or Email notification, on every occasion you install an App; it'll have its pre-decided configurations about just how it interacts with you (via notifications); nevertheless, you could easily change them, or forestall notifications from happening at all.

Lock display notifications on the iPad will come in two primary forms, banners and Notifications. Banners Appear near the top of the display screen, and Notifications which shows up at the guts of the lock screen. Probably the most glaring difference when working with your iPad is that banners vanish automatically after a couple of seconds; however, notification alert needs an action brought on (typically only a select) before they disappear.

Your iPad has notification middle; which shows you all the notifications you have obtained within the last 24 hours, that you haven't already taken action on - If you **OKAY** an alert for example, then it will not show in the notification middle.

Gain access to the iPad Notification Centre

Swipe down from the very best side of the iPad screen, and you will see two headings near the top of the screen notifications.

Follow the below process to gain access to the notification middle;

- "Today" gives you an accurate overview of your programs for your day - the weather, any calendar entries, and almost every other things you've gained access to. You might scroll to the low part and edit the Apps that could show information there.

- "Notifications" shows all information from the previous day that you have previously used. All Apps that generate notifications is seen in this list if you don't choose to exclude them (I will display for you ways to do this below).
- Tapping on any notification upon this list will need you to the little bit of information in the App that produced it - Tapping on the tweets notification, for example, would open up the tweets App which specific tweet.
- To clear notifications from confirmed App, select the x at the right-hand part of its name bar, then Tap **CLEAR**. There isn't any clear all function, lamentably, so you will want to achieve that for

each App in the list to vacant the list.

How to Change an App's Notification Preferences

Every App gets the same simple configurations as long as notifications are concerned.

- From the home display, find and select **SETTINGS**.
- In the primary settings menu, Tap **NOTIFICATIONS**.
- At the very top, you might see options showing the order where notifications are shown within the notification center. You can transform this by Tapping on this and choosing almost every other option. Below this, you will notice a summary of every one of the Apps on your iPad, and beneath their name, you will see their notification options.

To edit a credit card application options, Tap its Name.

- You may locate switches to regulate whether to allow Notifications if to show on Lock Display or in the Notification Centre and various options. You can additionally select if notifications should show as Banners or Notifications when the iPad is unlocked, or by no means, by Tapping the right choice.

While you're pleased with your configuration for the App, Tapping the blue arrow at the very top left-hand corner will lead you back to the menu, helping you to edit various other App's configurations.

Quick actions from Notifications

If you get a text message or email even while you’re utilizing your iPad, you’ll get a banner notification that shows near the top of the screen. You may get off this by swiping it upwards to continue using what you're doing. However, if you swipe it downwards, you'll access three other available choices.

Swiping downwards on the text notification will show the keyboard, letting you instantly answer the message and never have to change application. Swiping down on a contact notification gives you the decision of replying to or deleting the e-mail.

Around the lock display screen, you could additionally swipe from left on a contact notification to send it right to garbage without even unlocking your iPad.

How to Update the iPad Operating System (iOS Version)

Under normal circumstances, the operating-system of your iPad called **iOS** (***iPad Operating-System***) would look for updates itself, and also bring to your notice when the first is available by putting a red badge on the configurations icon. It will not set up itself though; you will do that!

To set up an update to iOS, follow the below steps:

- At the home screen, locate and select on **SETTINGS**.
- Inside the settings menu, Tap **GENERAL**.
- Tap on **COMPUTER SOFTWARE UPDATE**.

If the phone is current, you will see a message on-screen letting you know so. When there is any upgrade available, information for this will be shown on-screen together with a setup prompt as a means to begin the revise. You can likewise consent to new terms & conditions for the working device too!

Your phone may restart several time later during the

update process, and also you may not have the ability of utilizing it as it’s updating. So; it is best to set up an update on your phone when you understand you're not in all probability to need it within the timeframe the update would maintain progress!

CHAPTER 16

How to Extend iPad's Battery Strength

With every iPad release, one continuous point remains. The iPad is now faster, and the images get better every year. Nevertheless, the device works with the first 10 hours of battery life. However, also for those individuals that use their iPad 24hrs each day, it's still easy for it to perform. Moreover, there is nothing worse than looking to stream video from Netflix and then have that low battery message pop-up and interrupt your show.

Thankfully, there are a few tips you could utilize to keep iPad battery life and hold that from happening as often.

Concealed secrets that will change you into an iPad expert:

Right here's how you can get the best of your iPad's

battery life span:

- **Adjust the brightness:** The iPad comes with an automatic-brightness feature which facilitates the iPad predicated on the light quality within the area, but this program isn't always enough. Modifying the overall lighting could be the first solitary thing you can do to help ease out a little more from your battery consumption. You can transform the light by starting the iPad's configurations, choosing screen & view from the left-side menu and moving the brightness slider. The goal is to get it to a stage where it's nevertheless comfortable enough to learn, however nearly as shiny as the default establishing.

- **Switch off Bluetooth:** Most of us haven't any Bluetooth devices linked to the iPad, so all the

Bluetooth carrier does for all of us is waste charge of the iPad's battery life. When you have no Bluetooth devices connected, ensure Bluetooth is switched off. A brief way to turn the transfer for Bluetooth off is to open up the iPad manager - panel by swiping up from the backside of the screen.

- **Switch off Location services:** At precisely the same time as even the Wi-Fi version of the iPad will do a fantastic job of identifying its location, most people do not use the location service on our iPad as much as we utilize them on our iPad. Turning OFF GPS is a concise and clean manner to save lots of a little battery even while not quitting any feature. Also, remember if you need to use GPS, you may switch it **ON**. You can switch off location services in the iPad's configurations below privacy.

- **Turn off notification:** While notification is an excellent feature, it can drain a small amount of battery life because the tool assessments to check on if it needs to force a notification to the screen. If you want to do the most to optimize your battery life, you can turn drive notification off completely. You could likewise turn it off for specific apps, reducing all of the push notifications you obtain. You may switch off notification in configurations.

- **Fetch email less regularly:** With default configurations, the iPad will check out brand spanking new email each quarter-hour. Pushing this back to a half-hour or one hour can help your battery last much longer. Move to configurations, choose the email settings and select the "fetch new data" choice. This will enable you to set how often your iPad fetches email. There could even be a choice only to have check email manually.

- **Switch off 4G:** More often than not, we use the iPad at home, this means the use from it via our mobile data connection on. We utilize it at home exclusively. If you regularly end up low on battery, a good suggestion is to turn off your 4G data connection. This may protect it from draining any power when you are not using it.
- **Turn off background application refresh:** Background app refresh keeps your apps up to date by relaxing them even while the iPad is idle or as long as you're on various other apps. This may

drain a little extra battery strength. Get into configurations, choose General configurations and scroll down till you find "background app refresh." You can choose to turn off the service or certainly flip off specific apps you don't want to run in the background.

- **Discover apps consuming your battery:** Do you realize you could test thoroughly your iPad's battery usage? That is a fantastic manner to find what applications you're using the most and which apps can be consuming more than their expected percentage of your battery. You could test utilization within the iPad's configurations by choosing battery from the left-hand side menu.

- **Match iPad improvements:** It's continuously

important to keep iOS updated to the latest from Apple. Not merely would this help optimize the battery life of the iPad, it also ensures you're getting the latest security fix and patching any bugs that have popped up, which allows the iPad run efficiently.

- **Reduce Animation:** That is a technique to save a little of battery life and make the iPad show up a bit more reactive. The iPad's user interface consists of a few animations like glass windows zooming in and zooming out, and the parallax influence on icons, which makes them seem to hover over the backdrop picture. You can turn off those user interface effect by heading to **Settings**, Tapping **General Settings**, Tap **Display,** and getting to reduce animation to get it turned **Off**.

- **Buy a Smart Case:** The smart case can save battery life by placing the iPad into **Sleep mode** when you close the flap. It might not look like a great deal of conservation, however in case you aren't with the habit of striking the rest/wake button each time you have completed using the iPad; it could help offer you a supplementary five, ten or even quarter-hour extension of battery strength by the end of the day.

Chapter 17

Amazing iPad Tips all User Should Know

The iPad is an excellent tablet, yet most users have no idea about all the useful tips and shortcuts that makes life more facile with it. New iOS improvements are continuously adding cool new features, and that means you can be left outdated unless you know everything your iPad could probably do for you. We'll help you learn some of these critical features here.

- Find applications quickly

How will you find a specific application you installed on your iPad when you have lots of apps? Don't spend your time flipping through the displays; instead, use the iPad's Limelight search, which may be accessed by swiping down on the screen. Once you get accustomed to searching the iPad, you will not know how you'd be patience to do it some other way; you can even use this solution to search through your contact or even your email.

- Miss/skip the apostrophe when typing

The iPad's autocorrect will often get on your nerves, but at other times it really can be excellent. If you type a great deal, you'll without a doubt need to use the apostrophe regularly, mainly when you are typing in a contraction like "can't" or "won't"; do you realize you can miss the apostrophe? *The best iPad typing suggestion is using the autocorrect to improve "can not" to "can't" and "will not" to "won't."*

- Quick onscreen music controls

The iPad has buttons privately for changing the volume,

but think about skipping a song? You don't have to release the music application to miss a track. The iPad's control panel enables you to do things such as adjusting the lighting of the display screen, turn off/on the Bluetooth, and even reach the timer. *Slip your finger up from the bottom part of the display. You can pause, play, or forward or backwards.*

- Connect your iPad to your HDTV

You aren't limited by the iPad's screen if you are watching a movie or playing a casino game. You can even connect the iPad to an HDTV. The simplest way is through Apple Television, which facilitates Air-Play and enables you to wirelessly cast your iPad's display screen to your Television. But even though you have no desire for Apple TV, you can purchase an adapter to plug your iPad into the TV. *The very best solution is Apple's Digital*

AV Adapter; nevertheless, you can also get composite or component cables.

- Divide the safari browser in two

You may need a newer/latest iPad to utilize this tip. The iPad Air 2, iPad Mini 4, and iPad Pro or more modern tablets can start using a break-up view feature with the Safari internet browser. This splits the web browser into two home windows hand and hand, that allows you to see two websites at the same time. As the iPad requires a little elbow room because of this one, you must be keeping the iPad in scenery mode.

To enter break up view in the Safari internet browser, tap and contain the Webpages button: the button in the upper-right part of the Safari display that appears like a square within another square. When you click this button, you will notice all your open webpages, but when you possess your finger down onto it, a menu shows up that gives you the decision of opening divide view (if your iPad facilitates it), opening a new tab, or shutting all your Safari tabs.

If you are in a break-up view, this menu appears at the bottom of the screen. *To summarize divided view, do a similar thing: keep down the Web pages button to get the choice to combine all tabs.*

- <u>Use a custom keyboard</u>

Instead of skipping the apostrophe when typing, another option is installing a brand-new screen keyboard on your iPad; given that widgets are backed, you can use a custom keyboard. These keyboards come with many different advantages, like the capability to attract words by maintaining your finger pressed against the screen while moving from notice to letter, a method that sounds unusual but actually will save lots of time. You can use a third-party keyboard by downloading one from the App Store and turning it on in the iPad's keyboard configurations.

- Add applications to the home screen bottom level tray

The iPad includes four applications underneath the tray of the home screen, but do you realize you can include up to six applications to it? You can also take away the ones that are there by default and add your own.

Touch and hold an application icon until all the applications start shaking; this enables you to go through applications around by dragging them with your finger. To get an application into the bottom level tray, pull it down and drop it on the holder. You will see the other applications move to make room for this, and that tells you it is alright to drop it.

Pro Suggestion: You can drop entire folders into the bottom level tray too. If you have a couple of video games, you always want fast access, to simply put all of them in a folder and then drop the folder in this tray.

- Organize your applications with folders

While using Limelight to find apps is fantastic, typing within an app name every time you want to gain access to it might not be for you. Folders enable you to reach apps with a few taps or swipes; you should use folders to easily organize your iPad and individual apps into customized categories. The iPad will generate a default folder name that is usually a pretty good explanation of the applications it includes; nevertheless, you can rename the category to whatever you want.

To make a folder, *hold your finger down on an app's icon*

until all the apps start to tremble. Next, move the app together with another app, and the iPad will generate a folder made up of the apps. To include more applications to the folder, pull them over, and drop them on the recently created folder.

As stated earlier, you can move folders to underneath tray; you should use this to make a menu-system of types containing your preferred applications by dragging multiple folders to the holder. You can also arrange your iPad so that the majority of your applications are stored in folders lined over the bottom level tray as well as your most-used applications are on the first web page of the home screen.

- The iPad's virtual Touchpad can make you forget your mouse

Do you realize there is a virtual touchpad included in

your iPad? This touchpad might not be as effective as genuine, but it's close. You can use it any moment the on-screen keyboard shows up. Keep two fingertips down on the keyboard and move them around the screen; you will notice a triggered. As you move your fingertips around the display screen, the cursor will move with them. If you click and keep an item before moving your fingertips, you can also select text message this way, and you don't have to tap your fingertips on the real keyboard because of this to work. You can touch two fingertips anywhere on the display to activate the touchpad.

- Reboot the iPad

Do you realize you can solve many issues with the iPad by merely rebooting it? *Is your iPad operation slow? Reboot it. Will an application quit each time you start it? Reboot it.*

Unfortunately, you can confuse placing the iPad into suspend mode with a hard reboot. To necessarily give your iPad a brand-new start, you can reboot it by pursuing these quick steps:

1. *Press down the Rest/Wake button at the same time as pressing the home button*: maintain these for a couple of seconds. The iPad will go empty as it shuts down.

2. *Press down the Rest/Wake button again as it starts booting up*: When you start to see the Apple logo design show up, you can release the Rest/Wake button. The iPad's home display screen can look momentarily.

- <u>Save battery life by turning down display brightness</u>

An instant way to save your iPad's electric battery is to

lessen the display screen brightness. You can do this by locating the iPad's configurations and choosing Screen & Brightness from the left-side menu. (When you have a mature iPad, the choice may be called Brightness & Wallpaper.) You can move the slider to change the Brightness. Adjust the slider to improve the light of the display, the dimmer the screen; the less battery it uses.

- Disable in-app purchases

A very important factor every parent ought to know how to do is “switching off in-app buys on the iPad.” Usually, that seemingly free game may finish up, costing a massive amount of money; after your seven-year-old purchases, a couple of in-game costs might increase to $4.99 a touch. It's pretty easy to prevent this; first, you will need to *allow parental settings by locating your iPad's configurations and choosing General from the left-side menu. On this display, find Limitations. In the Limitations/Restriction menu, you will have to enable limitations, that may ask you for a four-digit passcode.*

Once you've enabled these parental handles, it is only a matter of scrolling listed below until you start to see the option for In-App Buys. When you glide this to the Off position, most applications won't even show the display screen for purchasing items, and the ones that do will be avoided from going right through with any transactions.

- Control your personal computer from your iPad

Want to consider things a step further? It is possible to control your personal computer from your iPad. This works on both Windows-based PCs and Macs. *You need to install software on your computer as well as an application on your iPad, but it is simple to manage it.* There are a bunch of excellent free software option you can get online, although if you intend on utilizing it extensively, you might like to go with paid tools.

CHAPTER 18

How to Fix iPad that won't Charge

If you're having troubles charging your iPad when it's linked to your personal computer, you are in good company. While your iPad or iPod device might not have a difficulty charging when connected to the USB slot on your pc, the iPad requires substantially higher power. This means that some USB slots, especially the ones on old computer systems, genuinely don't have the feature to get the duty performed.

How to see whether your iPad is charging

If the device has enough power to charge the iPad, a lightning bolt can be shown within the center of the battery meter at the very top right-hand corner of the iPad.

If it generally does not have enough capacity to charge the iPad, you might start to see the phrases "not charging" next to the battery meter.

Usually, the iPad can be charged using the computer if you positioned it in active mode. However, if the computer is set into sleep mode, the iPad will not charge.

The smooth and straightforward answer is to plug the iPad into a power outlet by using the adapter that is included with the iPad. That's also the quickest way to charge the iPad. Even personal computers that could effectively charge the iPad may not place out almost as much current as a standard charger. Some power strips additionally have USB ports to charge USB devices, which may be an incredible manner to charge the iPad.

Trouble charging the iPad when Connected to Power Outlet

First, be sure the iPad does not have a software problem by rebooting it. To achieve that, press down the sleep button on the top right-hand corner of the iPad. After some mere seconds, a crimson button can appear instructing you to slip it to the power of these devices.

Allow it shut down completely, and then click the power button down to power ON. You might start to see the apple brand logo appear at the center of the screen even while it boots up.

If the iPad still doesn't charge using the electrical outlet, you may have to check your cable or the adapter. You could discover if you have trouble with the cable by linking the iPad to your laptop with the cable.

If you start to see the lightning bolt on battery meter or

"not charging" next to the battery meter, you know the wire is working. If this is the case, absolutely buy a new adapter.

If the computer will not respond when you plug in the iPad, it is then not recognizing the iPad connected, meaning the issue is possibly with the cable.

In rare circumstances, while changing the adapter and the cable will not do just fine, you might have a hardware issue with the iPad. If so, you'll need to contact Apple for support. (If you live near an Apple Store, try getting in touch with the customer support. Apple Store staff can be quite accommodating.)

CHAPTER 19

The solution to iPad that falls inside water or damaged by Water

If catastrophe has struck your iPad, do not worry. Even if you immerse your iPad into a full tub of water; all hope is probably not lost. It is normal to presume water splashing on an iPad leading to electric shock, dark smoke, and damaged screen. Nonetheless, it can be harder than you think for water to reach that circuitry. Also, one of the top motives for iPad failing after being submerged in water is the electric battery turning out to be corroded, which will not occur immediately.

You will find two exceptional types of water hazard as it pertains to an iPad; therefore, there are two specific actions you have to take. The first problem is when spilling water on an iPad. This has similar hazards combined with the iPad being unintentionally sprayed with water. The next form of risk is the iPad being dipped or immersed into plenty of water in a shower, a pool, a lake, and many more.

How to proceed if you spilled water on your iPad

This is where you honestly desire you have an incredible accessory shielding your device. Trust it or not, the iPad is remarkably waterproof. The exterior part of the iPad is dominated by an aluminium framework, which gives water little chance to get into the iPad. Even the rims are improbable to permit any water through from the time water is spilt on the iPad to when you clean it clean.

This leaves a few parts of the device to save from problems such as the speakers, the headphone jack, the lightning connector, the volume buttons, the power button, and the home button.

When you have your iPad wrapped in a smart case or an identical cushty-healthy case, it is impossible for water to get into it. You need to carefully dry the iPad, noting whether or no more any water become pooled across the home button, and then cautiously take away the case and then wiping away any water, inspect the rims of the iPad for any water, paying the most interest to the inside of the iPad.

If the surface is dry and there's no water on the home button, you're likely fine. However, it's usually best to leave the iPad unused for about 24 to 48 hours only to be sure it's fine.

I've experienced an iPad survived the spillage of a full glass of water directly on the screen. It is significant how accurate the experience of the smart case can maintain times of non-permanent clumsiness.

If you weren't lucky to have your iPad protected during such incidence, you might have to adhere to the instructions for handling a completely submerged iPad. If you just got some water on the screen, so you realize it didn't get close to the control keys, especially the home button, or the sound system/loudspeaker or USB slot, you would be right wiping it down. However, if water went everywhere into the iPad, play it secure by assuming water made its way into the panel.

How to Fix iPad Submerged in Liquid or water

You might have noticed how an iPad or iPad or a few other devices can be fixed after been plunged into a box of rice and remains there for a day. The main advantage of this process, sagely advice time would help save a damp iPad.

Research by Gazelle relates how rice, oatmeal, and even silica gel packets may not be almost as absorbent as we might think. Moreover, a silica gel packet won't suck water through Aluminium.

If you have noticed how some crystallized types of kitty litter are much like silica gel, remember they're additionally no more than rice (or smaller!).

If you wish to be more secure, use silica gel packets.

They will not get caught of one's iPad and cause more problems.

After absolutely drying the outer area of the iPad with a very soft towel or fabric; the huge decision is if to turn ON or OFF the iPad. If the iPad continued to be ON and powered, this choice is more straightforward: switch it OFF by pressing down the Power button and then either sliding the button to power it OFF when brought ON or to press down the Power button before iPad shut down itself.

Recall, the iPad being in sleep mode isn't like the iPad being shut down. The different parts of the iPad are nonetheless running at precisely the same time as it is in sleep mode, and most severe, the iPad can awaken itself if you have a notification, text, Facetime call, etc.

However, if the iPad has already been in sleep mode, waking it up to shut it down may be worse than leaving it in droop/sleep mode. This depends on one main thing: the chance that something will be needed to awaken the display. This can be a scheduled appointment reminder; a smartphone call routed to the iPad, a message, a Facebook notification, etc. It would help if you chose to turn off the iPad to allow it dried out. If possible, move the cursor around in progress and wake the iPad up and immediately shut it down with the use of the Power button and the instructions mentioned above. In many cases, the probability of the iPad getting up may be impossible; in cases like this removing it from sleep mode is excellent.

Do's and Don'ts

- **Don't:** use a locks dryer or leave your iPad near a heater or use any form of warmness which you wouldn't usually blast on your arm for one hour. A higher amount of heat can damage an iPad.
- **Do:** Leave your iPad for at the least a day and ideally 48 hours. You will need to leave the iPad seated up with the home button at the lowest (bottom level). Gravity is your friend. If any water made it into the iPad, it likely managed to get in through the home button, lightning interface, or bottom loudspeakers. Leaving your iPad position

up for two times can help that dampness make its way from the iPad. When you have an iPad with four audio speakers like the iPad Pro, you could wait a day and then turn the iPad position for the next day. This may optimistically permit any water to drip out through the sound system (loudspeakers).

If you want to use silica gel packets, be sure the iPad is in the upright position. Gravity remains your best friend, so you should make sure it's working for you as well as the gel packets.

My iPad cannot Power ON after being left for hours

With a little luck, the mere truth that the iPad sat for a few days is enough for just any stray wetness in the iPad to evaporate. If the iPad does not **Power ON** or if it'll power on but has apparent problems with changing colours on the screen or it freezes immediately, you must take it to the nearest Apple Store or dispatch it to Apple. A common cause for water damage and mould to intervene with an iPad is the harm done to the battery, and a battery may be all you have to get it working once again.

You can discover an apple retail region by using this web page https://www.apple.com/retail you may even reach apple's tech support team at +1-800-676-2775.

CHAPTER 20

How to Connect iPad to your Television Wirelessly and with Cable

The iPad remains a great way to take pleasure in movies and TV, specifically when viewing on that beautiful 12.9-inch iPad Pro. This makes the iPad an unusual way to slice the wire and get rid of cable. Moreover, the newest TV app requires this step further, adding a centralized spot to control all of your loading apps. For everyone who've hunted through each application looking for which channels a specific film or show, we thank Apple!

However, how about watching your TV? If you'd opt for viewing on your large screen, don't get worried, it is on the other hand easy to get your iPad linked to your

Television. You can also get it done wirelessly! Plus, you can connect your headphones to any Television to get a non-public watching experience.

Connecting the iPad to your TV with Apple TV and Airplay

Apple TV is an excellent way of connecting your iPad to your Television. While it is expensive over other alternatives, it's the most reliable solution that is mobile. This implies you will keep your iPad on your lap and utilize it as a remote control while sending the screen to your Television. This is the best and quality solution for video games, where using a cord to hook up your iPad on your Television may be prescribing or restricting.

Apple TV uses Airplay to have a connection/interaction

with your iPad. Most loading apps use airplay and send full-display screen 1080p video to it. However, even applications that don't support Airplay or video out will continue to work via screen mirroring, which replicates your iPad's screen on your Television.

Various other bonus of Apple TV is the applications already installed on these devices. If you love ***Netflix, Hulu plus*** and ***crackle***, you certainly do not need to add your iPad to see loading video from these services. The applications run natively on Apple TV. **Apple TV** additionally works amazingly with the iPad and iPod, letting you stream video through Airplay or use your entertainment device's audio speakers to try out music.

Apple currently arrived with a fresh release of Apple TV that works at the same processor used for the iPad air. This helps it be lightning swift. It additionally helps a complete-blown version of the application store, which provides usage of even extra apps. Regrettably, the access-level charge is $149 as at the time of research. The freshest information is the classic (old) Apple TV still works just superb, allowing you to connect your iPad on your TV's display and is currently designed for around $69 as at the time of research.

Connecting the iPad Wirelessly without the utilization of Apple TV via Chromecast

In case you don't have to head the Apple TV path; however nevertheless need for connecting your iPad to your TV without any wires, **Google's Chromecast** can be an alternative solution. It comes with an extraordinarily easy setup process that uses your iPad to configure the Chromecast and get it connected to your Wi-Fi network, and when the whole lot is established and operating, you might connect the iPad's screen to your television - as long as the application you are using, works with Chromecast.

Also, that's the best restricting factor when compared with Apple TV: Chromecast support wants to be included in the app when compared with Apple television's Airplay, which fits with almost every application for the iPad.

Why use ***Chromecast***? For just one factor, it's much less costly. You could buy a Chromecast for as cheap as $30. It will work with both Google android and iOS devices; if you come with an android phone with your iPad, you might use chromecast with all of them. Moreover, with android, chromecast has a function much like Apple TV's screen mirroring.

Connecting the iPad to your High Definition TV through HDMI Cable

Apple's digital AV adapter is most likely the very best & most straight-ahead way to hook your iPad up to your High definition TV. This adapter lets you connect an HDMI cable from your iPad to your Television. This wire will send the video out to your Television. This means any application that helps video out will arrive in 1080p "HD" quality.

Moreover, preferably Apple TV, the digital AV adapter works with show mirroring, so even applications that do not support video away will screen up to your Television set.

Are you concerned about battery life?

The adapter additionally gives you access for connecting

a USB cable to your iPad that can source power to the unit and keep maintaining that battery from running low when you are binging on Seinfeld or doing other things. You could additionally stream your film collection from your laptop to your iPad via your High definition TV using home posting. This is a long notch process to finally change from DVD and Blu-ray to digital video without dropping the "to view it" on your substantial display TV. You can purchase the apple lightning digital AV adapter on amazon.

Consider: The lightning connector cannot work with the first iPad, iPad 2 or iPad 3. You may want to buy a digital AV adapter with a 30-pin connection for those old iPad models. This makes an Airplay solution like Apple Television even far better for these models.

Connecting an iPad through Composite/Component Cables

In case your TV does not support HDMI, or if you're genuinely working low on HDMI outputs on your HDTV, you can also choose to connect the iPad for your TV with composite or component cables. The amalgamated adapters break up the video to red, blue and green that provides a somewhat better picture, but amalgamated adapters are best designed for the old 30-pin adapters. Component adapters use the one 'yellowish' video wire well suited with the red and white audio cables, that's appropriate for almost all television sets.

The component and composite cables won't support the display mirroring mode on the iPad, so they could only use applications like Netflix and YouTube that support video out. Additionally, they flunk of 720p video;

therefore, the high-quality will never be as the digital AV adapter or Apple Television.

Unfortunately, these accessories might not be accessible to the more recent lightning connector, so; you'll need a lightning to 30-pin adapter.

Connecting the iPad with a VGA Adapter

Using Apple's VGA adapter, you could hook your iPad up to television outfitted with a VGA input, a pc monitor, a projector, and other display devices that supports VGA. That is amazing for screens. Many modern video screen units aid several display properties; you might even switch between your use of your display screen for your processing device and utilizing it for your iPad.

The VGA adapter will also support the screen mirroring mode. However, it generally does not transfer sound, and that means you will either pay attention through the iPad's built-in sound system or exterior speakers founded via the iPad's headphone jack.

If you're making programs on viewing through your TV, the HDMI adapter or the component cables will be the best solutions. However, if you intend on utilizing a laptop screen or want to use your iPad for big shows with a projector, the VGA adapter could be an excellent solution.

Do you realize you can view live Television on your iPad? There are many add-ons made to help you watch live Television on your iPad, access your cable stations and even your DVR from any room inside and while abroad through your computer data connect. Find out ways to watch TV from your iPad as you read further.

CHAPTER 21

How to Connect an iPad to a Wired Ethernet Port

The iPad was created to be considered a wireless device, and alas, it doesn't come with an Ethernet port allowing you to connect simultaneously to a router or network port. However, there are a few methods you can find and connect your iPad to an Ethernet network port or the back of your router.

A clean manner of executing that is to go through Wi-Fi certainly. If the number one goal is to connect your iPad into a network where there could be a slot available however no cellular network, you could utilize a transportable router and an Ethernet wire as a chance.

These pocket-sized routers can be of an enormous

solution because they don't require several different adapters to work. Plug in the cellular router and connect to the network. The ***ASUS*** transportable ***Wi-Fi router*** is approximately how big a debit card is and may switch a network port into a Wi-Fi hotspot. The ***Zyxel*** pocket trip router is similarly made to be ultra-portable.

Those routers will often have a short setup process that starts with locating the router in the Wi-Fi settings of your iPad. Once connected, you will proceed through a setup means that enables you to develop a secure connection.

In the event you must go wired, you may use the new charging **USB 3 adapter**. Apple identifies this adapter as a *"camera connection package,"* nonetheless it can connect any compatible USB device to the iPad. It would help if you used this adapter for connecting a wired pad, MIDI devices and, of course USB-to-Ethernet cables.

You can find two massive variations between your new charge to USB 3 adapter and the old camera connection kit. First, the latest adapter employs USB 3, which allows for considerably quicker transfer speed. Second of all, the new adapter includes a charging slot for a reason for plugging into a power outlet. This enables you to charge your iPad when you use the adapter, and significantly, it allows the use of the adapter to provide power.

Ethernet cables require the ability to work.

This solution works satisfactorily by using Apple's latest USB to Ethernet adapter with model number *MC704LL/A or the later version after this publication*. There could be a few issues using the old USB to Ethernet adapter or the utilization of third-party adapters, but, you might be in a position to get a tutorial to get different cables to work correctly.

You should first connect the Power to USB 3 adapter to the iPad. **Next**, plug the adapter into a wall structure power outlet by using the charging wall plug adapter that arrived with your iPad. Once you've supplied power, connect the USB to Ethernet adapter and then connect it to the network using an Ethernet wire.

How to Connect to Ethernet utilizing a powered USB Hub

The first trouble getting the iPad connected into Ethernet is the necessity for power. The iPad will not source power whether it's operating on battery power, so the completely new Charging to USB 3 adapter helps handle that problem. However; when you don't have the classic Charging to USB adapter? Alternatively, imagine if your USB to Ethernet adapter fails with the new camera connection package?

The answer: put in a powered USB port to the mix.

It has to be noted that workaround can be wonky for insufficient a better phrase. If the whole lot is made in the correct order, it must work, but because of the fact this system entails doing something the iPad wasn't prepared to do, it is not always guaranteed to work consistently.

You might want a powered USB hub as well as the USB camera connection kit and the USB to Ethernet adapter. Know that these may additionally grow to be costing more than searching for a tour-sized Wi-Fi router.

Once you've got the entirety, connecting your iPad is not

too difficult. Earlier than you begin, switch off Wi-Fi for the proper measure.

You may also need to ensure the USB hub is connected to a wall outlet. Again, the procedure won't work with no centre providing power.

- First of all, hook the Charging-to-USB connection kit to the iPad. (When you have an iPad with the 30-pin connection, you might want the 30-pin USB adapter).
- Next, sign up for the iPad to the USB interface utilizing a USB wire. Attach the USB-to-Ethernet adapter to the USB slot, and, connect the Ethernet adapter to a router or network interface by using an Ethernet wire.
- If you face any issue, attempt rebooting the iPad and start the process again.

CHAPTER 22

How to Fix an iPad that won't Upgrade iOS Version

Have you got an application that won't update or a whole new application that is swept up at the center of the download? That is fairly common, and there are a variety of reasons why an app could easily get swept up at the downloading section. More often than not it's both an authentication trouble, meaning the application store is having a hard time determining who you are, or there could be a problem with some other app or a few information that the iPad wants to download and the application is just waiting in line.

Moreover, on a few unusual occasions, the iPad forgets about the app. However do not be concerned, if you have this issue, the steps below would fix and restore it.

Tap the Application as though to release it

We shall begin with the iPad by exiting the app. *So how exactly does this happen?* Sometimes, a download will stall out credited to an unhealthy connection or similar reason, so make sure you have a good connection with the internet. You might make the iPad to begin downloading the application again by obviously looking to start the app. When you Tap on an app that is in the *'waiting to download'* stage, the iPad will attempt to download it.

Look for pending downloads in iTunes

If Tapping on the application didn't rectify the problem, you may determine if there could be anything in-line beforehand of the app. A typical trouble that triggers apps to avoid upgrading is while a piece of music, article, movie, or a similar little bit of content get stuck when downloading. If you're a typical consumer of **iBook's**, have a look at any books you are currently downloading and **Tap** them to ensure that they continue downloading. You must additionally go directly to the iTunes store application for your iPad to check for pending downloads. Inside the iTunes app, select the **purchased Tabs**. Movies would be sorted by the newest. Music and Television shows have a *"current buys"* hyperlink at the very top you can use to check for just any pending downloading. Again, surely **Tap** the thing to inform your iPad to keep downloading it.

Reboot the iPad

After checking the most typical motives for an application never to update or download completely, it's time to go with the utmost famous troubleshooting steps:

Reboot these devices

Consider, it is not sufficient to essentially droop/suspend the tool and wake it up again. If you wish to supply the iPad a complete refresh, you may want to power off these devices by pressing down the rest/wake button for numerous secs and following instructions on the screen. When it's entirely powered off, you could start it by pressing the rest/wake button again. This system will deliver the iPad a smooth beginning and comes with an inclination to resolve many troubles.

Download a whole new app

It is normal for the iPad to get hung within the center of the authentication method. This might keep the iPad from seeking to authenticate with the iTunes store again, which will freeze all downloads to your iPad. The most effective way to solve this difficulty is to download a whole new app, which allows you to make the iPad to authenticate again. Try to decide on a free application and set it up on the iPad. Once it installs, locate the precise application that was trapped to find out if it begins downloading.

Delete the application and download it again

Note that this task should not be attempted if the application saves records that you would like to keep, with a note-taking use or a pulling app.

A lot of those applications are saved on the cloud, which means that it is secure to delete, however, if you have any uncertainties, you will need to move this task by. If nothing else happens, spend some time working and not worried about the information you've created within the app, you might connect your iPad to your personal computer and check out iTunes with your laptop to peer if the data files can be copied to your home computer.

If the application will not keep figures or if the statistics are saved to the cloud much like applications like Evernote, delete the application and re-download it from the application store. You might need to sign-in afresh into the app when it's downloaded.

Logging Out of your Apple ID

If going right through the authentication technique by downloading an application doesn't appear to work, simply logging out and logging in is probably going to do the secret magic. You may find *Sign-in* from your **Apple ID** by clicking on it to open up the iPad's configurations, selecting the iTunes & application stores at the left-hand side menu and Tapping where it shows your **Apple ID**.

This may deliver up a popup menu to help you log out. Once you are authorized out, Sign-in again into the **Apple ID** and try starting the application once more.

Restart your Wireless router

While uncommon, it's more practical for your router to be the foundation of the problem. This is not always intentional. Your router is not mad at you or whatever; however, as it comes with an integrated firewall and managing multiple devices, it could get an impression combined up sometimes. Attempt running down the router and leave it off for a complete minute before turning the router back again on. It usually takes a router a few moments to **Power ON** and gets linked to the internet again. When all the lights keep coming back **ON**, attempt putting your password on it using your iPad and connect with the application to look if the download process has begun. Recall, you'll be without access to the internet throughout this process, so if there are certainly others within the home that will be utilizing the web connection, you must tell them.

Reset all settings

The next trick inside our arsenal is to reset the iPad's settings. Usually do not worry, this may not completely clean your iPad, however since it clears configurations, you will eventually lose any configurations previously custom-designed. You will also want to register back to websites that often synchronizes your accounts configurations. However, aside from clearing out your configurations, this technique won't wipe all of your apps, files, movies, and information or data.

To reset your configurations, cross into the iPad's

Settings/configurations and choose **General** from the left-hand side menu. Next, scroll every all the way down and select **reset**. With this screen, choose reset all configurations.

That is one of the utmost conventional therapies for an application that is caught during an update or an application that may not download completely; however since it can alternate any custom settings again to default, this task is saved for next-to-last option.

Reset your iPad

If clearing out the settings don't appear to be effectively working, it is time to take a little extra drastic action. The ultimate trick is to reset the iPad. This wipes out your apps, information, music, and many more. However, you could additionally bring these back (restore) from a backup.

The straightforward procedure is similar to getting a completely new iPad or iPad. When it is wiped, you will feel the same method you experienced when you initially received these devices, such as putting your password into iCloud and selecting if to restore from a back-up. The result is that you ought to have the capability to complete this system without losing all of your apps, songs, films or data. If you've ever improved your iPad or iPad to a new tool, you'll be familiar with the outcome.

However, you must think about considering if updating the application you want to update is worth it. You will be better off to delete the application and move ahead.

You could reset your device by entering settings, choosing General, simply clicking reset and choosing "*erase all content and settings*."

CHAPTER 23

How to Watch TV on Your iPad

Among the exquisite issues about the iPad is for you to use the Tablet for viewing TV. There are a few appropriate alternatives that support you to view Television on your iPad, and that means you don't have to miss your preferred show or any big game.

Many of these devices functions by intercepting its transmission out of your wire package and then broadcasting it through your wi-fi network, that enables you to access your shows from any place in the home even though on the run via your iPad's data connection.

However, one thrilling option converts the iPad into a lightweight TV, and if you do not want to invest the money on expensive add-ons or accessories, sometimes an application is all you have to.

Sling TV

The war to cut the cord has been waging for a couple of years now, but when sling TV surfaced, the tide officially started to be towards the cable companies. Sling TV is not a tool that throws your current day cable indicators to your Television or an antenna that accumulates channels for your iPad. It's an online cable company, due to this, they do precisely the same simple process that your wire organization will, without strolling wire to your dwelling. Instead, you stream your channels throughout the internet. Sling provides you with the usage of popular stations like ESPN, AMC, TNT, TBS as well as others. You can additionally sign up to HBO, Epix, and further channel packages.

The very best part is that sling TV needs to forget about

hardware to work. The terrible part is that it will not broadcast local stations.

TiVo Stream

If you're not thinking about cutting the cable, TiVo could be the best answer for pressing your television sign to your iPad. Sadly, it might be the most luxurious, at least beforehand charges. You'll either want to include a TiVo stream device on your existing TiVo set up, or you will require the TiVo Roamio Plus, which include both TiVo Roamio for essential TiVo supplier and the TiVo motion for loading throughout the internet.

However, while getting set up on TiVo is more expensive, you might forestall renting an HD Dvd movie recorder from your wire issuer, which may be helping you save money.

The huge bonus to TiVo stream is the ability to watch recorded shows in addition to concert events. The flow fundamentally turns your iPad into a TiVo player, which means you get access to the whole great deal documented on your TiVo's DVR.

Slingbox Slingplayer

No more to be confused with Sling TV, Slingbox's Slingplayer works via intercepting the TV signal away off your wire box and "slinging" it across your home network. The Slingplayer software transforms one's body into a good deal that enables you to stream it to your iPad across both Wi-Fi as well as your iPad's 3G/4G/5G data connection. Using the Slingplayer app, you can listen to other stations watching any television screen that you can view at home. You can also gain access to your DVR watching recorded shows.

Beyond being truly a very significant way to view remotely, Slingplayer is also an incredible solution for people who want usage of the television in virtually any room inside without connecting wire or springing for two televisions. One drawback would be that the iPad application should be bought individually and provides an incredible chunk onto the charge of these devices.

@TV Plus

Belkin is the newest manufacturer to leap into the market place, and their @TV is priced among the volcano glide and the Slingbox Slingplayer. It offers all precisely the same primary functions of both products, which include the ability to pause and record live TV.

One feature oddly lacking from @TV can be an HDMI connection. @TV best uses an element or amalgamated cables instead of the better acceptable HDMI.

Cable Television / Network Apps

A chance to starting up hardware on your entertainment device or plugging an antenna into the iPad is downloading applications from your Wire Company or primary networks.

Many major companies like COX TV, Time Warner Cable, and DirectTV offer applications for the iPad, and that means you can watch TV, even though not all support the full total range of TV provided through your subscription rather than all provision ability to stream over 3G/4G/5G.

You could additionally access top-class articles via apps, with a few restrictions based on your issuer. HBO, Cinemax, Showtime and Starz all have applications that utilize a few providers.

Beyond reputable applications from your wire issuer or top quality channels, there are a few remarkable applications for streaming films and TV. The two most popular choices are **Netflix**, which gives an excellent selection of movies and Television for a comparatively low subscription charge, and **Hulu Plus**, which doesn't have precisely the same movie series but offers a few Television shows still within the present-day season.

Crackle is likewise an excellent choice for loading movies and will not require any membership charges.

CHAPTER 24

How to Fix a Slow iPad

Is your iPad working slowly? Might it get bogged down after a couple of hours? At the same time as this is extra, not uncommon with old iPads that don't have the control power of the new iPad Air and iPad Pro Tablets, even the latest iPad can impede. A couple of multiple reasons just why an iPad could also begin operating slow, such as an application having troubles or a sluggish web connection. Thankfully, that is generally easy to revive.

- **First: stop all of your new apps**

One common reason behind an iPad to start chugging is a problem with the application itself instead of the iPad. If you enjoy a form that is working slower than usual, it could sound reasonable to go through the home button to close the application and re-launch it. However, pressing the home button wouldn't normally close out the app. It suspends the app, which mostly keeps it freezing in the backdground.

Some applications even continue steadily to run in the backdrop mode. Those are usually applications that stream music like Pandora, Spotify, or the melody app that is included with the iPad.

If the hassle is specifically with an individual app, we'll need to stop from it using the duty display. This may correctly close the application down and purge it from memory space, permitting you to release a 'fresh' version from it. Please discover that you might lose unsaved functions by exiting the app. If it's currently working on an objective, it could be better to make the application finishes the duty before proceeding.

Within the job screen, it is an excellent concept to have a

summary of any applications that are taking part in music. It's not likely they may be causing a headache, or even if the application is loading the track from the web, it will not expend enough of your bandwidth to rely upon. However, last from the app won't hurt, and maybe sure the application isn't impacting something.

To shut the program, you will need to constitute a summary of all apps that will be operating in the background.

- **Double-click the home button in the bottom of your iPad.**

When you press it two times in quick succession, your most up to date applications are shown as cascading home windows across the screen. You might navigate via this screen by swiping from left-to-right or right-to-left. The windows can have a related application icon above it.

***To close an app:

- Keep the finger down at the home display window.
- Without lifting your finger from the screen, swipe in the direction of the very best of the screen.

This gesture resembles "flicking***" the application from the iPad. Remember: you Tap the application windowpane, not the application icon.

Reboot the iPad

Closing the applications may not continuously do just fine. Within this example, rebooting the iPad is the product quality recourse. This will flush from memory and offer your iPad a fresh start.

NB: many humans believe the iPad forces down as the rest/wake button on the top right-hand corner of the iPad is pressed down or as the flap of their smart cover or bright case is closed, but this places the iPad in droop/suspend setting.

To reboot the iPad:

- Keep down the rest/wake button until instructions appear; letting you know to glide a button to power from the iPad.
- When you slip the button, the tablet will turn off, and the iPad's screen should go dark.

Wait around several seconds and then start the iPad up pressing down the rest/wake button once again. You'll first start to see the Apple brand logo design on the screen as well as your iPad need. Your iPad must run extra fast but, if it begins bogging down, retain in mind the apps that are running at that time. Once in a while, a single application can purpose the iPad to execute poorly.

Is your iPad still walking slower than you want?

Check your wireless connection

It could not be your iPad that is working sluggishly. It could be your mobile network.

You can attempt the internet speed of your Wi-Fi network by using an application like ***Ookla's Speedtest***. This application will send information to a remote server and then dispatches records back to the iPad, looking into each send and download rates of speed.

The usual Wi-Fi network in the U.S. Gets around 12 megabits-per-second (Mbps), though it is unusual to see rates of speed of 25+ Mbps. You possibly might not see plenty of the slowdown using your connection until it receives around 6 Mbps or significantly less. It is across the amount of bandwidth it requires to stream films and video.

If you're experiencing a headache with your Wi-Fi

connection, try moving nearer to your router. If the pace increases, you might look at improving your Wi-Fi range. That's common in more prominent structures, but even a little home could have troubles.

Ensure you're updated to the latest version of iOS

iOS is the operating device working on the iPad. At precisely the same time as the best revise sometimes will positively sluggish the iPad down a little, it's always a great idea to perform the latest operating-system on device. Not handiest will this make sure that you have the most up to date efficiency tweaks, it additionally warranties you have present fixes for just about any protection problems.

Setup an ad-blocker

If you're mainly seeing a decline when surfing the

internet in the safari internet browser, however, your web velocity is not sluggish, with the ability to be more an indicator of which webpages you're surfing than the iPad itself.

The more significant advertisements with a web page, the much longer it does take to load. Moreover, if some of those advertisements stall out, you might be left anticipating the web page to pop-up.

One fashion to this is to set up an advertisement blocker. Those widgets beautify the safari web browser by using disallowing advertisements to weight on the internet website. They make each for more straightforward reading and faster launching. Sites such as this one generate income from adverts, which means this is a balance you have to fight.

Flip off background app refresh

Background app refresh let applications to refresh their content even when you are not utilizing them. In this manner, Facebook might Tap base and retrieve articles in your post wall, or an information app may also fetch the newest articles.

However, this runs on your processing rate and your web connection, so that it can make the iPad to perform just a little slower. This usually isn't the theory cause, but if you frequently find the iPad working slow (and if the electric battery drains quickly), you must flip off background app refresh.

Showing off background application refresh:

- Head to your iPad's configurations.
- Choose **General** from the left-hand navigation menu.

- Tap the background application refresh.
- Tap the on/off slider near the top of the screen.

If you are nevertheless experiencing progressive speeds, there is certainly yet another factor you can do.

Clear STORAGE SPACE

If you're operating desperately low on space for storage, clearing up a little more room for the iPad can on occasion, improve efficiency. This will be achieved by deleting applications that you haven't used for quite a long time, particularly video games you don't play anymore.

It's clean to see which applications are employing the most space on your iPad:

1. Head to **Settings.**
2. Select **General** from the left-hand navigation

menu.

3. Tap **storage space & iCloud usage**.

4. Tap **Manage Storage Space** (under the very best storage program). This may demonstrate which apps are using up the utmost storage.

You can additionally increase safari when you delete your cookies and internet background, although this might purpose you to log back to any websites that have saved your login records.

LET US HEAR FROM YOU!

We love testimonies. We love to hear what your opinions are from the information you have read in this author's book. Please share your experience and lesson learnt from this book with us, as we look forward to always serve you better.

Also, *please consider giving this book a* ***good review*** *on the product page where you purchased, and check out other books by the same author and other book recommendations by our publishing house.*

We also invite you to check out our website at http://www.engolee.com/bookdeals and consider joining our newsletter growing list of hundreds of thousands of book lovers, which we send **Free and Discounted books** to weekly, with great tips, testimonies and resourceful content from the various authors that connect with us.

Feel free to drop us your suggestions, opinions, testimonies etc. at info@engolee.com We would be eager to listen and support you, and also work on any advice given to us.

CPSIA information can be obtained
at www.ICGtesting.com
Printed in the USA
LVHW010253070121
675552LV00001B/32